Ben Stacy Jerrik (Ed.)

Lund Municipality

AF307137

Ben Stacy Jerrik (Ed.)

Lund Municipality

Urban areas in Sweden, Municipalities of Sweden , Skåne County, Dalby Söderskog National Park

Part Press

Imprint

Permission is granted to copy, distribute and/or modify this document under the terms of the GNU Free Documentation License, Version 1.2 or any later version published by the Free Software Foundation; with no Invariant Sections, with the Front-Cover Texts, and with the Back- Cover Texts. A copy of the license is included in the section entitled "GNU Free Documentation License".

All parts of this book are extracted from Wikipedia, the free encyclopedia (www.wikipedia.org).

You can get detailed informations about the authors of this collection of articles at the end of this book. The editors (Ed.) of this book are no authors. They have not modified or extended the original texts.

Pictures published in this book can be under different licences than the GNU Free Documentation License. You can get detailed informations about the authors and licences of pictures at the end of this book.

The content of this book was generated collaboratively by volunteers. Please be advised that nothing found here has necessarily been reviewed by people with the expertise required to provide you with complete, accurate or reliable information. Some information in this book maybe misleading or wrong. The Publisher does not guarantee the validity of the information found here. If you need specific advice (f.e. in fields of medical, legal, financial, or risk management questions) please contact a professional who is licensed or knowledgeable in that area.

Any brand names and product names mentioned in this book are subject to trademark, brand or patent protection and are trademarks or registered trademarks of their respective holders. The use of brand names, product names, common names, trade names, product descriptions etc. even without a particular marking in this works is in no way to be construed to mean that such names may be regarded as unrestricted in respect of trademark and brand protection legislation and could thus be used by anyone.

Cover image: www.ingimage.com
Concerning the licence of the cover image please contact ingimage.

Publisher:
Part Press is a trademark of
International Book Market Service Ltd., 17 Rue Meldrum, Beau Bassin, 1713-01 Mauritius
Email: info@bookmarketservice.com
Website: www.bookmarketservice.com

Published in 2012

Printed in: U.S.A., U.K., Germany. This book was not produced in Mauritius.

ISBN: 978-613-6-28326-5

Contents

Articles

Urban_areas_in_Sweden	1
Lund_Municipality	4
Municipalities_of_Sweden	6
Skåne_County	10
Dalby_Söderskog_National_Park	15
List_of_national_parks_of_Sweden	16
Dalby,_Lund	27
Viborg_Municipality	29
Hamar	30
Porvoo	41
Dalvík	46
León,_Nicaragua	48

References

Article Sources and Contributors	53
Image Sources, Licenses and Contributors	54

Urban_areas_in_Sweden

Urban area is a common English translation of the Swedish term *tätort*. The official term in English, used by Statistics Sweden, is, however, **locality**. There are 1,940 localities in Sweden (2005). They could be compared with *census-designated places* in the United States.

A *tätort* in Sweden has a minimum of 200 inhabitants, and may be a city, town or larger village. Urban areas referred to as cities/towns (*stad*) for statistical purposes have a minimum of 10,000 inhabitants.[1] However, since 1971, the term "stad" is no longer in use as a judicial concept in Sweden.

History

Up until the beginning of the 20th century only the cities were regarded as urban areas. The built-up area and the municipal entity were normally almost congruent. Urbanization and industrialization created, however, many new settlements without formal city status. New suburbs grew up just outside city limits, being *de facto* urban but *de jure* rural. This was of course a statistical problem. The census of 1910 introduced the concept of "densely populated localities in the countryside". The term *tätort* (literally "dense place") was introduced in 1930. The municipal amalgamations placed more and more rural areas within city municipalities, which was the other side of the same problem. The administrative boundaries were in fact not suitable for defining rural and urban populations. From 1950 rural and urban areas had to be separated even within city limits, as e.g. the huge wilderness around Kiruna had been declared a "city" in 1948. From 1965 only *non-administrative localities* are counted, independently of municipal and county borders. In 1971 *city* was abolished as a type of municipality.

Map of Sweden showing all urban areas (cities and towns) with a population of more than 20 000.

Terminology

Urban areas in the meaning of *tätort* are defined independently on the division into counties and municipalities, and are defined solely according to population density. In practice, most references in Sweden are to municipalities, not specifically to towns or cities, which complicates international comparisons. Most municipalities contain many localities (up to 26 in Kristianstad Municipality), but some localities are, on the other hand, multimunicipal [2]; Stockholm urban area is spread over 11 municipalities.

When comparing the population of different cities, the urban area (*tätort*) population is to prefer ahead of the population of the municipality. The population of e.g. Stockholm should be accounted as ~1.2 million rather than the ~800,000 of the municipality, and Lund rather ~75,000 than ~110,000.

Swedish definitions

Terms used for statistical purposes

- *Tätort* (= urban area or locality) is the central concept used in statistics. The definition is agreed upon in the Nordic countries: A *tätort* is any village, town or city with a population of at least 200 for which the contiguous built-up area meet the criterion that houses are not more than 200 meters apart when discounting rivers, parks, roads, etc.[3] Every fifth year (2000, 2005 etc.) the localities are revised and new population figures released.
- *Småort* (= minor locality) is a rural locality with 50–199 inhabitants in a contiguous built-up area with no more than 150 meters between houses.[4] The concept is rarely used outside the field of statistics, where it is used for settlements just below the limit defined for *tätort*.
- *Centralort* (= central locality) is mostly used in the meaning municipal seat or municipal center of service, commerce and administration for an area.

Popular and traditional terms

- *Storstad* (= literally "large city"; metropolitan area) is a term usually reserved for Sweden's three largest cities: Stockholm, Göteborg (Gothenburg) and Malmö. Statistics Sweden uses the term metropolitan area ("storstadsområde") for these three cities and their immediate surroundings and municipalities.[5]
- *Stad* (= town or city) is in a context of statistics restricted to urban areas with a population greater than 10,000.[1] Judicially, the term *stad* is obsolete since 1971, and is now mostly used describing localities which formerly were chartered towns. The statistical category "large town" used by Statistics Sweden include municipalities with more than 90,000 inhabitants within a 30 km radius from the municipality centre.[6] There is also a category *medelstor stad* "middle large town".
- *Köping* (= market town) was also abolished as an official term in 1971 in governmental and statistical contexts, and is only rarely kept in use by laymen, although it has survived as part of the names of several smaller towns. The meaning was a locality with an intermediary legal status below that of a town.
- *Municipalsamhälle* (= municipal community) was a term in use between 1875 and 1971, but it is no longer used outside of historical contexts. In 1863, Sweden was divided into 2,500 municipalities, whereof 89 were towns, 8 were market towns (köpingar) and the rest rural municipalities ("landskommuner"). A "municipalsamhälle" was an administrative centre for one or several rural municipalities, with special regulations and privileges in common with towns. The term became obsolete in 1971 when the different types of municipalities were abandoned and a standard form for all municipalities was introduced.
- *Samhälle* (= community) is a common concept used by for urban areas that are intermediary in size between a town and a village. The term "samhälle" is also used in Swedish to denote "society", "community" or "state". (Compare: Gemeinschaft and Gesellschaft.) A *samhälle* does not necessarily meet the criteria for the current *tätort* — or even *småort* concept.
- *By* (= village and hamlet) is a traditional term but may in colloquial use refer to a suburb or town of considerable size. If at all used in the context of statistics, it must be assumed that the size of a *by* is smaller than that of a *småort*. (NB! Not to be confused with the same word in Danish and Norwegian, where it means town, while a village is called *landsby*.)

Seasonal areas and suburbs

- *Fritidshusområde* (= seasonal area) is in statistical context an area with less than 50 permanent inhabitants but at least 50 houses (in practice: weekend cottages/summer houses) meeting the criterion that they are not more than 150 metres apart. About a third of Sweden's "second homes" are located in such areas. The term belongs also to everyday usage, although less strictly defined.
- *Förstad* and *förort* (= suburb) are much used terms with a somewhat negative connotation.

Statistics

Data are computed by Statistics Sweden every five years. The latest data are as of 31 December 2005. Then the total population of the urban areas (or localities) in Sweden was 7,631,952 on an area of 5,286.23 km², which gives an average population density of 1,444/km².

- 84% of the Swedish population lives in localities (i.e. in *tätorter*)
- 50% lives in the 64 largest urban areas
- A third lives in the 15 largest urban areas
- A quarter lives in the 5 largest
- The largest and most populous urban area is Stockholm

See also

- List of urban areas in Sweden
- Largest urban areas of the European Union
- Geography of Sweden
- List of municipalities of Sweden
- List of cities in Sweden
- Largest urban areas of Norway

References

[1] Statistics Sweden. Be 16 SM 9601 (http://www.scb.se/statistik/MI/MI0810/2000I02/MI38SM9601.pdf), Tätorter 1995, p. 2: "Towns (localities with more than 10,000 inhabitants)".

[2] http://w41.scb.se/templates/Publikation____186288.asp

[3] Statistics Sweden. Be 16 SM 9602 (http://w41.scb.se/statistik/MI/MI0810/2000I02/MI38SM9602.pdf), Småorter 1995, Befolkningskoncentrationer i glesbygd, English summary, p. 3: "Locality (having at least 200 inhabitants) = urban area". Retrieved 2 December 2007.

[4] Statistics Sweden. Agglomerations in rural areas 1995 (http://w41.scb.se/statistik/MI/MI0810/2000I02/MI38SM9602.pdf), Småorter 1995 Befolkningskoncentrationer i glesbygd. Retrieved 2 December 2007.

[5] Statistics Sweden. Population in the metropolitan areas on Dec. 31, 2002 and 2003 (http://www.scb.se/statistik/BE/BE0101/2003A01/BE0101_2003A01_BR_05_BE76SA0401.pdf), SCB Befolkningsstatistik del 1-2, 2003. Retrieved 2 December 2007.

[6] Statistics Sweden. Press release (http://www.scb.se/Pages/PressArchive____259760.aspx?PressReleaseID=169883), Household budget survey (HBS), 2006-06-01 Nr 2006:079A. Retrieved 2 December 2007.

External links

- *Tätort*-statistics (http://www.scb.se/Pages/Product____12991.aspx) from Statistics Sweden

Lund_Municipality

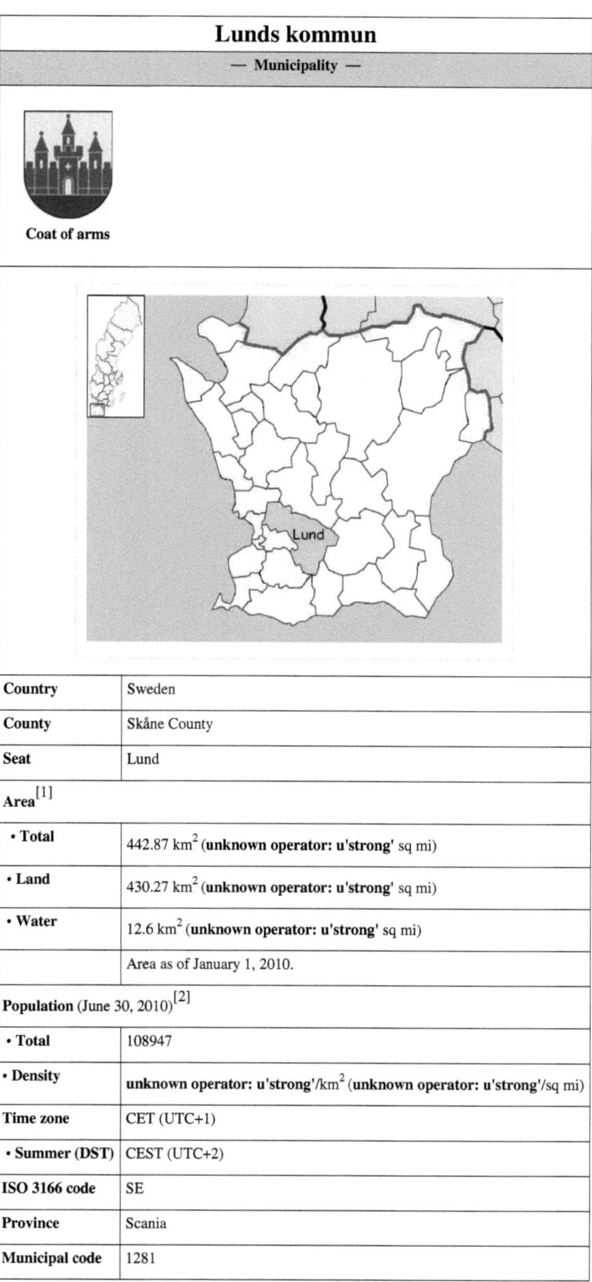

Lunds kommun	
— Municipality —	
Coat of arms	
Country	Sweden
County	Skåne County
Seat	Lund
Area[1]	
• **Total**	442.87 km^2 (**unknown operator: u'strong'** sq mi)
• **Land**	430.27 km^2 (**unknown operator: u'strong'** sq mi)
• **Water**	12.6 km^2 (**unknown operator: u'strong'** sq mi)
	Area as of January 1, 2010.
Population (June 30, 2010)[2]	
• **Total**	108947
• **Density**	**unknown operator: u'strong'**/km^2 (**unknown operator: u'strong'**/sq mi)
Time zone	CET (UTC+1)
• **Summer (DST)**	CEST (UTC+2)
ISO 3166 code	SE
Province	Scania
Municipal code	1281

Website	www.lund.se [3]

Density is calculated using land area only.

Lund Municipality (*Lunds kommun*) is a municipality in Skåne County, southern Sweden. Its seat is located in the city of Lund.

As most municipalities in Sweden, the territory of Lund Municipality consists of a lot of former local government units, united in a series of amalgamations. The number of original entities (as of 1863) is 22. At the time of the nation-wide municipal reform of 1952 the number had been reduced to six. In 1967 the rural municipality *Torn* (itself created in 1952 was added to Lund. The *City of Lund* was made a unitary municipality in 1971 and amalgamated with *Dalby, Genarp, Södra Sandby* and *Veberöd* in 1974 completing the process.

Dalby Söderskog, one of Sweden's national parks, is located within the municipality between the Dalby and Lund.

Localities

There are nine urban areas (or localities) in Lund Municipality.

In the table the urban areas are listed according to the size of the population as of December 31, 2010. The municipal seat is in bold characters.

#	Locality	Population
1	**Lund**	82,800
2	Södra Sandby	6,136
3	Dalby	5,708
4	Veberöd	4,062
5	Genarp	2,892
6	Stångby	1,218
7	Idala	738
8	Torna Hällestad	584
9	Revingeby	545

International relations

Twin towns — sister cities

Lund has a sister city in each of the Nordic countries, as well as in other countries.[4]

- Viborg Municipality, Denmark
- Hamar, Norway
- Porvoo, Finland
- Dalvík, Iceland
- Nevers, France
- León, Nicaragua
- Greifswald, Germany
- Zabrze, Poland

References

- Statistics Sweden [5]

[1] "Statistiska centralbyrån den 1 januari 2010" (http://www.scb.se/Statistik/MI/MI0802/2010A01/mi0802tab3_2010.xls) (in Swedish)
(Microsoft Excel). Statistics Sweden. . Retrieved 2010-08-21.
[2] "SCB, Befolkningsstatistik 30 juni 2010" (http://www.scb.se/Pages/TableAndChart____244147.aspx) (in Swedish). Statistics Sweden. .
Retrieved 2010-08-19.
[3] http://www.lund.se
[4] Lund Municipality homepage, twin cities (http://www.lund.se/templates/Page____2731.aspx)
[5] http://www.scb.se

External links

- Lund Municipality (http://www.lund.se/) - Official site

Municipalities_of_Sweden

Sweden's municipal borders

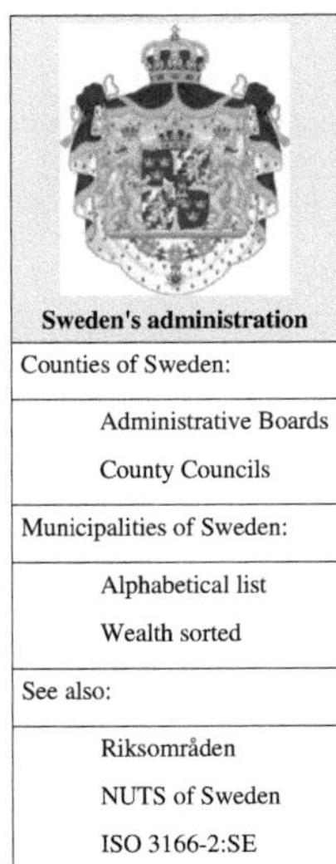

Sweden's administration
Counties of Sweden:
Administrative Boards County Councils
Municipalities of Sweden:
Alphabetical list Wealth sorted
See also:
Riksområden NUTS of Sweden ISO 3166-2:SE

The **municipalities of Sweden** (*kommun*) are the local government entities of Sweden. The current 290 municipalities are organized into 21 counties (*län*). The municipal governments are responsible for large portion of local services like schools, emergency services and city planning.

Foundation

The basic regulation of Swedish municipalities can be found in the Local Government Act of 1991. It specifies several responsibilities for the municipalities, and provides outlines for local government, such as the process for electing the municipal assembly. It also regulates a process (*laglighetsprövning*, "legality trial") through which any citizen can appeal the decisions of the local government to a county court.

Municipal government in Sweden is similar to city commission government and cabinet-style council government. A legislative municipal assembly *(kommunfullmäktige)* of between 31 and 101 members (always an uneven number) is elected from party-list proportional representation at municipal elections, held every four years in conjunction with the national parliamentary elections. The assembly in turn appoints a municipal executive committee *(kommunstyrelse)* from its members. The executive committee is headed by its chairman, (Swedish: *kommunstyrelsens ordförande*). The chairman is often referred as Municipal Commissioner (Swedish: *kommunalråd*).

History

The first local government acts were implemented on January 1, 1863. There were two acts, one for the cities and one for the countryside. The total number of municipalities was about 2,500. The rural municipalities were based on the old parishes (*socknar*) and the then 89 cities/towns (*städer*) (which is the same in Swedish) were based on the old chartered cities. There was also a third type, *köping* or market town. The status of these was somewhere between the rural municipalities and the cities. There were only eight of them in 1863, rising to a peak of 96 in 1959.

Up until 1930, when the total number of municipalities reached its peak (2,532 entities), there were more partitions than amalgamations.

In 1943 more than 500 of Sweden's municipalities had fewer than 500 inhabitants, and the *1943 års kommunindelningskommitté* ("Municipal subdivision commission of 1943") proposed that the number of rural municipalities should be drastically reduced.

After years of preparations the first of the two nation-wide municipal reforms of the 20th century was implemented in 1952. The number of rural municipalities was reduced from 2,281 to 816. The cities (by then 133) were not affected.

Rather soon it was established that the reform of 1952 was not radical enough. A new commission, *1959 års indelningssakkunniga* ("Subdivision experts of 1959") concluded that the next municipal reform should create new larger mixed rural/urban municipalities.

The Parliament of Sweden (*Riksdagen*) decided in 1962 that the new reform should be implemented on a voluntary basis. The process started in January 1964, when all municipalities were grouped in 282 *kommunblock*("municipal blocks"). The co-operation within the blocks should ultimately lead to amalgamations. The target year was 1971, when all municipalities should be of uniform type and all the remaining formal differences in government and privileges between cities and rural municipalities should be abolished.[1]

The amalgamations within the "blocks" started in 1965 and more were accomplished in 1967 and 1969, when the number of municipalities dropped from 1006 to 848. The Riksdag, however, found the amalgamation process too slow, and decided to speed it up by ending the voluntary aspect. In 1971 the unitary municipality (*kommun*) was introduced and the number of entities went down to 464; three years later it was 278. In one case (Svedala Municipality) the process was not accomplished until 1977.

Most of the municipalities were soon consolidated, but in some cases the antagonism within the new unities was so strong that it led to "divorces". The total number of municipalities has today risen to 290.

The question of whether a new municipality will be created is at the discretion of the central Swedish government. It is recommended that the lower limit of a new municipality shall be 5,000 inhabitants.

Some municipalities still use the term "City" (Swedish: *stad*) when referring to themselves, a practice adopted by the largest and most urban municipalities Stockholm, Gothenburg and Malmö. 13 municipalities altogether, some of them including considerable rural areas, have made this choice, which is unofficial and has no effect on the administrative status of the municipality. The practice can, however, create some confusion as the term *stad* nowadays normally refers to a larger built-up area and not to an administrative entity.

Geographical boundaries

The municipalities in Sweden cover the entire territory of the nation. Unlike the USA or Canada, there are no unincorporated areas. The municipalities in the north cover large areas of sparsely populated land. Kiruna, at 19 446 km², is sometimes held to be the world's largest "city" by area, although places like La Tuque, Quebec (28 421 km², official style *Ville*), Wood Buffalo, Alberta (63 343 km², official style "regional municipality") and the City of Kalgoorlie-Boulder in Western Australia (95 575 km²) are larger. (By comparison, the total area of the state of Lebanon is 10 452 km².) At any rate, several northern municipalities are larger than many counties in the more densely populated southern part of the country.

Sub-division

The municipalities are also divided into a total of 2 512 parishes, or *församlingar* (2000). These have traditionally been a subdivision of the Church of Sweden, but still have importance as districts for census and elections. Many of the parishes still correspond to the original *socknar*, but there have been a lot of partitions and amalgamations throughout the years.

Duties

According to law, the municipalities are responsible for:

- Childcare and pre-school
- Primary and secondary schools
- Social service
- Elderly care
- Support to people with disabilities
- Health and environmental issues
- Emergency services (not policing, which is the responsibility of the central government)
- Urban planning
- Sanitation (waste, sewage)

Many municipalities in addition have services like lesiure activities for youths and housing services to make them attractive in getting residents.[2]

See also

- List of municipalities of Sweden
- List of Swedish municipalities by wealth
- Local federation

References

[1] "Indelning i kommuner och landsting" (http://www.regeringen.se/sb/d/1906/a/12261) (in Swedish). Regeringen.se. . Retrieved 2008-09-20.
[2] "Levels of local democracy in Sweden" (http://www.skl.se/artikel.asp?A=48690&C=6393). Swedish Association of Local Authorities and Regions. . Retrieved 2008-09-25.

External links

- Swedish Association of Local Authorities and Regions (http://www.skl.se/startpage_en.asp?C=6390)
- The Local Government Act in English translation (http://www.regeringen.se/content/1/c4/06/96/34cb7541.pdf)
- Swedish Government (http://www.sweden.gov.se/) – Official site

Skåne_County

<table>
<tr><td colspan="2" align="center">Skåne County
Skåne län</td></tr>
<tr><td colspan="2" align="center">— County of Sweden —</td></tr>
<tr><td colspan="2" align="center">Seal</td></tr>
<tr><td>Country</td><td>Sweden</td></tr>
<tr><td>Capital</td><td>Malmö</td></tr>
<tr><td>Government</td><td></td></tr>
<tr><td>• Governor</td><td>Göran Tunhammar</td></tr>
<tr><td>• Council</td><td>Skåne Regional Council</td></tr>
<tr><td>Area[1]</td><td></td></tr>
<tr><td>• Total</td><td>11027 km^2 (unknown operator: u'strong' sq mi)</td></tr>
<tr><td colspan="2">Population (March 31 2011)[2]</td></tr>
<tr><td>• Total</td><td>1251213</td></tr>
<tr><td>• Density</td><td>unknown operator: u'strong'/km^2 (unknown operator: u'strong'/sq mi)</td></tr>
<tr><td>Time zone</td><td>CET (UTC+1)</td></tr>
<tr><td>• Summer (DST)</td><td>CEST (UTC+2)</td></tr>
<tr><td>GDP/ Nominal</td><td>SEK 278,254 million (2004)</td></tr>
<tr><td>GDP per capita</td><td>SEK 331,700 (2010)</td></tr>
<tr><td>NUTS Region</td><td>SE224</td></tr>
</table>

Skåne County (*Skåne län*), also known as **Scania County** in English, is the southernmost administrative county or *län*, of Sweden, basically corresponding to the traditional province Scania (which in Swedish is called Skåne). It borders the counties of Halland, Kronoberg and Blekinge. The seat of residence for the Skåne Governor is the town of Malmö. The headquarters of Region Skåne is the town of Kristianstad.[3]

The present county was created in 1997 when Kristianstad County and Malmöhus County were merged, and it covers around 3% of Sweden's total area, but its population of 1,250,000 comprises 13% of Sweden's total population.

Counties before 1997.

Endonym and exonym

When the new county was established in 1997, it was decided to name it after the historical province Scania, or *Skåne*, as it is called in Swedish, so it was given the name *Skåne län*. The most common name used in English is *Skåne County*[4] [5] [6], even if the form *Scania County* is not uncommon.[7] [8] [9] [10] It can not be considered wrong to use either form in English.

Heraldry

The coat of arms for Skåne County is the same as for the province of Scania only with the tinctures reversed and the crown, beak and tongue of the Griffin in the same color. When the arms is shown with a royal crown it represents the County Administrative Board, which is the regional presence of (royal) government authority. Blazon: "Gules, a Griffin's head erased Or, crowned and armed the same".

Provinces

Skåne County is the administrative equivalent of the province of Scania, but it also includes an insignificant part of the province of Halland.

Administration

The seat of residence for the Governor or *Landshövding* is the town of Malmö. The County Administrative Board is a Government Agency headed by a Governor. See List of Skåne Governors.

The two former administrative counties of the province of Scania shown on the map, Kristianstad County and Malmöhus County, which were established in 1719 were merged together in 1997, forming the present county with boundaries that are almost identical to the boundaries of the province.

County council

Skåne Regional Council (*Region Skåne*) is an evolved County Council, which was established in 1999 when the County Councils of the former counties were amalgamated. Its main responsibilities are for the public healthcare system and public transport. In addition, it has for a trial period assumed certain tasks from the County Administrative Board.

Its county or regional assembly is the highest political body in the region and its members are elected by the electorate,[11] as opposed to the county administrative board, that guards the national interests in the county under the chairmanship of the county governor (*landshövding* in Swedish).

Municipalities

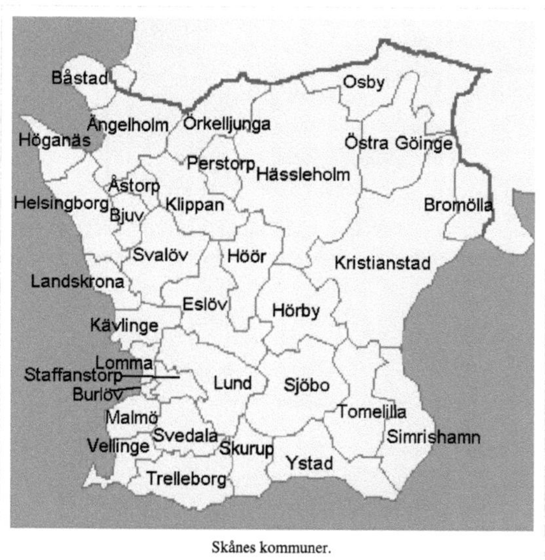

Skånes kommuner.

Skåne County is subdivided into 33 municipalities[12] (*kommuner* in Swedish), the largest being Malmö Municipality (301,000 inhabitants), Helsingborg Municipality (130,000), Lund Municipality (111,000 inhabitants) and Kristianstad Municipality (80,000 inhabitants). The municipalities have municipal governments, similar to city commissions, and are further divided into parishes. The parish division is traditionally used by the Church of Sweden, but also serves as a divisioning measure for Swedish census and elections.

- Bjuv
- Bromölla
- Burlöv
- Båstad
- Eslöv
- Helsingborg
- Hässleholm
- Höganäs
- Hörby
- Höör
- Klippan
- Kristianstad
- Kävlinge
- Landskrona
- Lomma
- Lund
- Malmö
- Osby
- Perstorp
- Simrishamn
- Sjöbo
- Skurup
- Staffanstorp
- Svalöv
- Svedala
- Tomelilla
- Trelleborg
- Vellinge
- Ystad
- Åstorp
- Ängelholm
- Örkelljunga
- Östra Göinge

Electoral districts

The county is divided into four parliamentary constituences or electoral districts, electing 45 of the 349 members of the *Riksdag*, the Parliament of Sweden. Each district is made up of one or more municipalities.

The political parties are represented by the following number of MPs from Skåne County after the Swedish general election, 2010:

- Moderate Party 16
- Swedish Social Democratic Party 12
- Liberal People's Party 4
- Green Party 4
- Sweden Democrats 4
- Centre Party 2
- Christian Democrats 2
- Left Party 1

The electoral districts of Skåne County for national parliamentary elections

Localities in order of size

The ten most populous localities of Skåne County as defined by Statistics Sweden 2005:

#	Locality	Population
1	**Malmö**	258,020
2	Helsingborg	91 457
3	Lund	76 188
4	Kristianstad	33 083
5	Landskrona	28 670
6	Trelleborg	25 643
7	Ängelholm	22 537
8	Hässleholm	22 548
9	Ystad	17 286
10	Eslöv	16 551

Transport

The motorway built between Malmö and Lund in 1953 was the first motorway in Sweden. With the opening of the Oresund Bridge between Malmö and Copenhagen (the longest combined road and rail bridge in Europe) in 2000, the Swedish motorways were linked with European route E20 in Denmark, and the two countries' railway systems were physically connected. Before the bridge was built there were train ferries operated between Helsingborg and Helsingør. There are also train ferries to and from Germany and Poland.

Scania has three major public airports, Malmö Airport, Ängelholm-Helsingborg Airport and Kristianstad Airport. One of the oldest airports in the world still in use is located in Scania, namely Ljungbyhed Airport, in operation since 1910. Starting in 1926, the Swedish Air Force used the airport for flight training, and up until the military school was moved to the nearby Ängelholm F10 Wing in 1997, the airport was extremely busy. In the late 1980s, it was Sweden's busiest airport, with a record high of more than 1,400 take-offs and landings per day.[13]

The major ports of Scania are Trelleborg, Copenhagen Malmö Port and Helsingborg Harbour. Ferry connections across the Baltic Sea operate from several smaller ports as well.

References

[1] http://www.sna.se/webbatlas/lan/skane.html Sveriges Nationalatlas. Retrieved 2 April 2008

[2] "Kvartal 1 2011" (http://www.scb.se/Pages/TableAndChart____228187.aspx). Statistics Sweden. .

[3] About Region Skåne (http://www.skane.se/templates/Page.aspx?id=54724). Region Skåne. Retrieved 10 April 2008.

[4] http://www.lansstyrelsen.se/skane/Om_Lansstyrelsen/In+English/

[5] http://www.sna.se/webatlas/county/skane.html

[6] http://www.polisen.se/en/English/Contact-the-Swedish-Police/

[7] Lund University (http://www.lucram.lu.se/board)

[8] Swedish Environmental Protection Agency (http://www.naturvardsverket.se/Documents/publikationer/620-5458-9.pdf)

[9] Nordicnet (http://www.nordicnet.net/city/Malmo/)

[10] Airport Nav Finder (http://www.airportnavfinder.com/airport/ESMS/)

[11] Region Skåne. Democracy-Increased autonomy (http://www.skane.se/templates/Page.aspx?id=43835). Official site. Retrieved 24 August 2007.

[12] Region Skåne. Municipalities in Skåne (http://www.skane.se/templates/Page.aspx?id=56606). Official site. Retrieved 24 August 2007.

[13] Ljungbyhed airport - ESTL (http://web.archive.org/web/20050907012811/http://www.tfhs.lu.se/english/skolan/6.html). Fact sheet created by Lund University School of Aviation (http://www.lusa.lu.se). Retrieved 22 January 2007.

External links

- Skåne County Administrative Board (http://www.m.lst.se/)
- Skåne Regional Council (http://www.skane.se/)

Dalby_Söderskog_National_Park

Dalby Söderskog National Park	
IUCN category II (national park)	

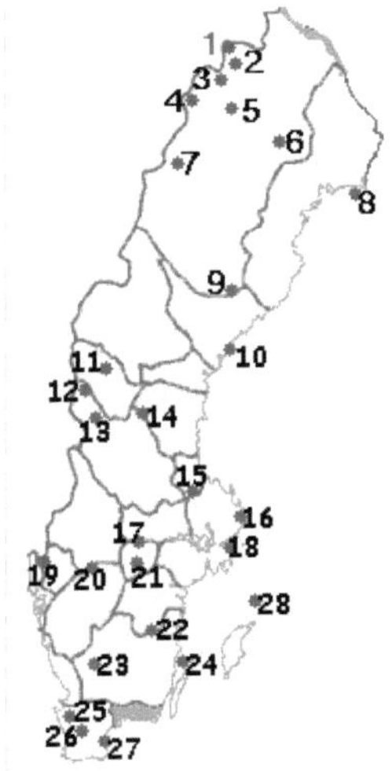

Location	Skåne County, Sweden
Nearest city	Lund
Coordinates	55°40′N 13°19′E
Area	0.36 km^2 (**unknown operator: u'strong' sq mi**)[1]

Dalby Söderskog is a small national park in the province of Scania in southern Sweden, situated in the municipality of Lund, near Dalby. It has an area of 0.36 km² and consists of broadleaf forest. It was established in 1918, when it was thought to be a unique remnant of primeval forest. In fact, the area has previously been used for pasture. The ground contains much limestone and chalk, which makes the flora rich. In particular, there are many spring flowers. An earth bank of unknown origin, possibly the ruins of an ancient fort, surrounds parts of the park. The national park encompasses the southern part of Dalby Hage, while the northern parts (Dalby Norreskog and Hästhagen) form a nature reserve.

Dalby Söderskog is number 26 on the map

References

[1] "Dalby Söderskog National Park" (http://www.naturvardsverket.se/en/In-English/Start/ Enjoying-nature/National-parks-and-other-places-worth-visiting/National-Parks-in-Sweden/ Dalby-Soderskog-National-Park/). Naturvårdsverket. . Retrieved 2009-02-26.

External links

- Dalby Söderskog National Park (http://www.naturvardsverket.se/en/ In-English/Start/Enjoying-nature/ National-parks-and-other-places-worth-visiting/National-Parks-in-Sweden/Dalby-Soderskog-National-Park/) from the Swedish Environmental Protection Agency

List_of_national_parks_of_Sweden

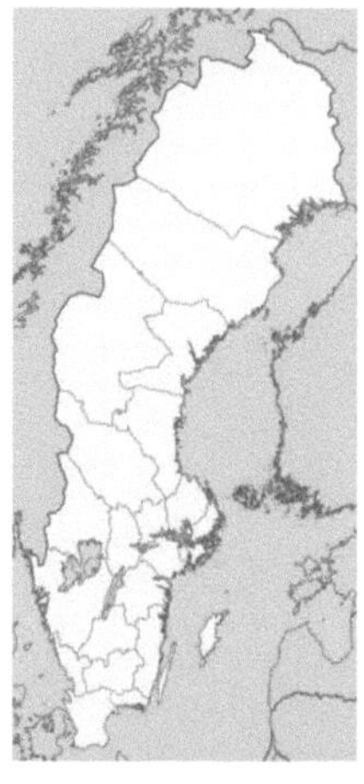

Abisko

Björnlandet

Blå Jungfrun

Dalby Söderskog

Djurö

Fulufjället

Färnebofjärden

Garphyttan

Gotska Sandön

Hamra

Haparanda Archipelago

Koster-
havet

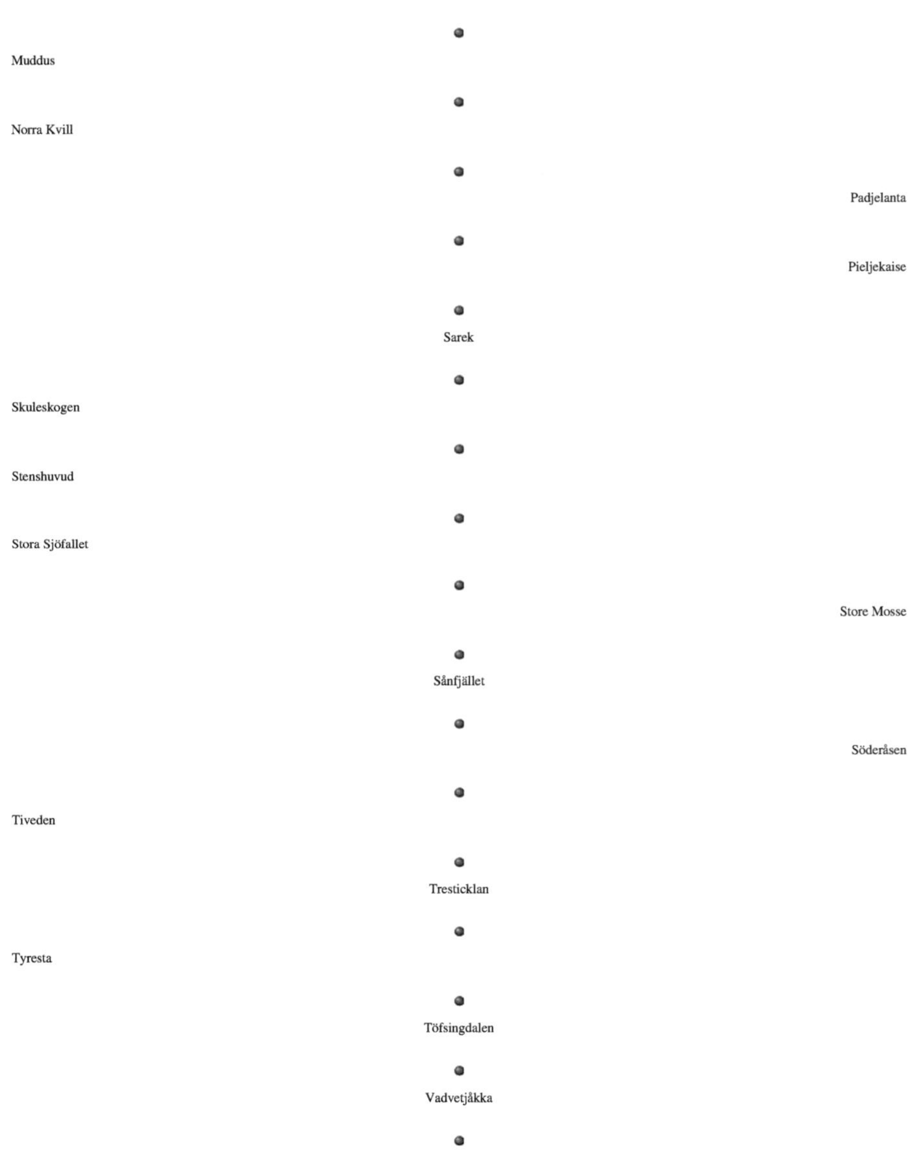

National parks of Sweden (clickable map)

National parks of Sweden are managed by the Swedish Environmental Protection Agency (EPA) (Swedish: *Naturvårdsverket*) and owned by the state. The goal of the national park service is to create a system of protected areas that represent all the distinct natural regions of the country.[1] In 1909, Sweden became the first country in Europe to establish such parks when nine were opened following the Parliament of Sweden's (Swedish: *Riksdagen*) passing of a law on national parks that year. This was followed by the establishment of seven parks between 1918 and 1962 and thirteen between 1982 and 2009, with the latest being Kosterhavet National Park.[1] There are currently 29 national parks in Sweden, comprising a total area of 731,589 hectares (1,807,796 acres);[2] six more are scheduled to open by 2013.[3]

According to the EPA, Swedish national parks must represent unique landscape types and be effectively protected and used for research, recreation, and tourism without damaging nature.[4] Mountain terrain dominates approximatively 90% of the parks' combined area. The reason for this is the extensive mountain areas taken up by the large northern parks—Sarek National Park and Padjelanta National Park each cover approximately 200,000 hectares (490,000 acres).[5] [6] Many of the northern parks are part of the Laponian area, one of Sweden's UNESCO World Heritage Sites due to its preserved natural landscape and habitat for the native reindeer-herding Sami people.[7] The southernmost parks—Söderåsen National Park, Dalby Söderskog National Park and Stenshuvud National Park—are covered with broadleaf forest and together cover approximately 2000 ha (**unknown operator: u'strong'** acres).[8] [9] [10] Fulufjället National Park is part of PAN Parks,[11] a network founded by the World Wildlife Fund (WWF) to provide better long-term conservation and tourism management of European national parks.[12]

National parks

This along with * indicates that the national park is part of a World Heritage Site

This along with ** indicates that the national park is a Protected Area Network Park

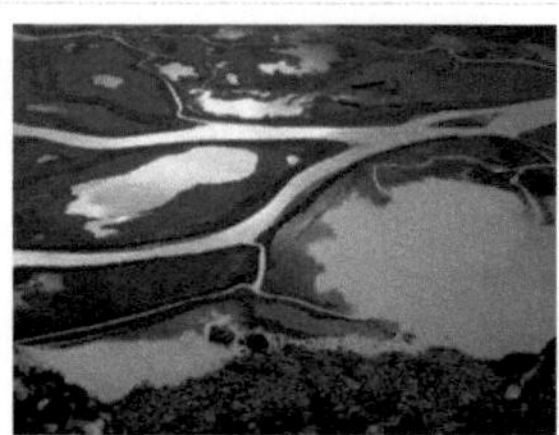

A river delta near Sarek National Park

Pierikpakte Mountain, part of Sarek National Park

Skäralid River, part of Söderåsen National Park

Abisko National Park was established in 1909.

Stream just over the tree line in Stora Sjöfallet National Park.

The forest in Dalby Söderskog National Park

The hill Stenshuvud in Stenshuvud National Park

Gotska Sandön National Park is one of Sweden's oldest national parks.

Name	Location[13]	Area[13]	Established[13]	Description
Abisko National Park	Norrbotten County	7700 ha (**unknown operator: u'strong'** acres)	1909	The park is composed of valleys framed by mountain ranges in the south and west and Scandinavia's largest alpine lake, Torneträsk, in the north.[14]
Ängsö National Park	Stockholm County	168 ha (**unknown operator: u'strong'** acres)	1909	Ängsö is an island in the Stockholm archipelago. The park is known for its "ancient farm landscape in the archipelago environment, the spring flowers, and the varied bird life".[15]
Björnlandet National Park	Västerbotten County	1100 ha (**unknown operator: u'strong'** acres)	1991	Björnlandet's geography is distinguished by its large virgin forest and mountain terrain with steep ravines and cliffs. The park features traces of several forest fires.[16]

Blå Jungfrun National Park	Kalmar County	198 ha (**unknown operator: u'strong'** acres)	1926	Blå Jungfrun is an island in the Baltic Sea dominated by clefts and hollows in the north and forest in the south.[17]
Dalby Söderskog National Park	Skåne County	36 ha (**unknown operator: u'strong'** acres)	1918	Deciduous forest surrounded by a 56 m (184 feet) wide earth bank that takes up a large part of the park.[9]
Djurö National Park	Västra Götaland County	2400 ha (**unknown operator: u'strong'** acres)	1991	Djurö National Park consists of an archipelago with about 30 islands in Sweden's biggest lake, Vänern.[18]
Färnebofjärden National Park	Dalarna, Gävleborg, Uppsala, and Västmanland counties	10100 ha (**unknown operator: u'strong'** acres)	1998	Dalälven River passes through the park and the uneven shoreline encloses over 200 islands and islets.[19]
Fulufjället National Park**	Dalarna County	38500 ha (**unknown operator: u'strong'** acres)	2002	The park consists mainly of bare mountain heights, and heaths that are unique in the Swedish mountains.[11]
Garphyttan National Park	Örebro County	111 ha (**unknown operator: u'strong'** acres)	1909	Garphyttan National Park consists of landscape altered by humans through agriculture and forestry, such as meadows and deciduous forest.[20]
Gotska Sandön National Park	Gotland County	4490 ha (**unknown operator: u'strong'** acres)	1909	Gotska Sandön is an island composed of sand. Its scenery is dominated by beaches, dunes, and pine forests.[21]
Hamra National Park	Gävleborg County	28 ha (**unknown operator: u'strong'** acres)	1909	Hamra National Park contains two low moraine hills covered with virgin forest and large rock boulders.[22]
Haparanda Archipelago National Park	Norrbotten County	6000 ha (**unknown operator: u'strong'** acres)	1995	Located in the northern part of the Gulf of Bothnia, the park is composed of low islands with wide sandy beaches.[23]
Kosterhavet National Park	Västra Götaland County	38878 ha (**unknown operator: u'strong'** acres)	2009	Kosterhavet National Park is the first national marine park of Sweden and was inaugurated in September 2009. It consists of the sea and shores around the Koster Islands, however excluding the islands themselves.[24] [25] [26] [27]
Muddus National Park*	Norrbotten County	49340 ha (**unknown operator: u'strong'** acres)	1942	Muddus National Park is home of deep ravines and primeval forests. Sweden's oldest pine tree is located in the park.[28]
Norra Kvill National Park	Kalmar County	114 ha (**unknown operator: u'strong'** acres)	1927	Norra Kvill is an ancient forest with tall pine trees that are over 350 years old. Three lakes are situated in the park: Stora Idegölen, Lilla Idegölen and Dalskärret.[29]
Padjelanta National Park*	Norrbotten County	198400 ha (**unknown operator: u'strong'** acres)	1962	The park, which borders Norway in the west, is primarily composed of a flat and open landscape that surrounds the two lakes Vastenjávrre and Virihávrre.[6]
Pieljekaise National Park	Norrbotten County	15340 ha (**unknown operator: u'strong'** acres)	1909	Pieljekaise National Park is composed of birch forest, mountain terrain, and several lakes. The park is named after Pieljekaise Mountain, a landmark in the area.[30]

Park	County	Area	Established	Description
Sånfjället National Park	Jämtland County	10300 ha (**unknown operator: u'strong'** acres)	1909	The park is named after the 1,278 m (4,192 ft) high mountain Sånfjället. The mountainous area is intersected by streaming lakes and a forest area.[31]
Sarek National Park*	Norrbotten County	197000 ha (**unknown operator: u'strong'** acres)	1909	The park features an alpine landscape with high peaks and narrow valleys. More than 100 glaciers are found in the park, and several mountains are over 2,000 m (6561 ft) high.[5]
Skuleskogen National Park*	Västernorrland County	2360 ha (**unknown operator: u'strong'** acres)	1984	Skuleskogen National Park is composed of ancient forest, high mountains, and sea coast. The mountain peaks are covered with pine forest and are separated by valleys formed by the sea and ice sheets.[32]
Söderåsen National Park	Skåne County	1625 ha (**unknown operator: u'strong'** acres)	2001	The park features an especially contoured landscape with up to 90 m (300 ft) deep ravines. The valleys are covered with broadleaf forest, mostly beech.[8]
Stenshuvud National Park	Skåne County	390 ha (**unknown operator: u'strong'** acres)	1986	Stenshuvud is a hill that faces the Baltic Sea. Because the surrounding landscape is relatively flat, it can be seen from a great distance and has been used by seafarers as an aid to navigation at sea. Most of the area is covered with broadleaf forest.[10]
Stora Sjöfallet National Park*	Norrbotten County	127800 ha (**unknown operator: u'strong'** acres)	1909	The park's northern portions lie in the Scandinavian Mountains, home to some of Sweden's highest peaks. The lower hills in the park's southern part are covered with forest.[33]
Store Mosse National Park	Jönköping County	7850 ha (**unknown operator: u'strong'** acres)	1989	Store Mosse National Park is the home of the largest bog area in southern Sweden. The lake Kävsjön, containing many species of birds, is located within the park.[34]
Tiveden National Park	Örebro County and Västra Götaland counties	1350 ha (**unknown operator: u'strong'** acres)	1983	Tiveden National Park is a part of the large Tiveden forest. The park is situated in the most inaccessible part of the forest. The landscape is mountainous and stony.[35]
Töfsingdalen National Park	Dalarna County	1615 ha (**unknown operator: u'strong'** acres)	1930	Töfsingdalen National Park consists of two mountain ridges separated by a valley covered with fields and virgin forest.[36]
Tresticklan National Park	Västra Götaland County	2897 ha (**unknown operator: u'strong'** acres)	1996	This park contains a rift valley landscape and is one of the few remaining areas of pristine forest in southern Scandinavia.[37]
Tyresta National Park	Stockholm County	2000 ha (**unknown operator: u'strong'** acres)	1993	Tyresta is a gorge landscape with stony slopes. The park, covered with pine forrest, is one of the largest virgin forests in Sweden.[38]
Vadvetjåkka National Park	Norrbotten County	2630 ha (**unknown operator: u'strong'** acres)	1920	Located in a mountain region north-west of Lake Torneträsk, Vadvetjåkka National Park is the northernmost national park in Sweden. The park is named after Vadvetjåkka Mountain, which is located within the park.[39]

Future national parks

In 2008, after investigations and interviews with the participating counties, the Swedish Environmental Protection Agency laid down a plan to establish 13 new national parks in the near future. According to the plan, seven of the parks will be established between 2009 and 2013, the first being Kosterhavet National Park which was inaugurated in September 2009. It is currently unknown when the six remaining parks will be established.[3]

Sweden's highest mountain Kebnekaise will be part of a national park sometime between 2009 and 2013.

The Sylan mountain range will be part of the Vålådalen-Sylarna National Park.

Name	Location[40]	Area[40]	Date of establishment[40]
Bästeträsk National Park	Gotland County	5000 ha (unknown operator: u'strong' acres)	2009–2013
Blaikfjället National Park	Västerbotten County	40000 ha (unknown operator: u'strong' acres)	2009–2013
Kebnekaise National Park	Norrbotten County	65000 ha (unknown operator: u'strong' acres)	2009–2013
Tavvavuoma National Park	Norrbotten County	40000 ha (unknown operator: u'strong' acres)	2009–2013
Vålådalen-Sylarna National Park	Jämtland County	230000 ha (unknown operator: u'strong' acres)	2009–2013
Västra Åsnen National Park	Kronoberg County	2000 ha (unknown operator: u'strong' acres)	2009–2013
Nämdöskärgården National Park	Stockholm County	14000 ha (unknown operator: u'strong' acres)	TBA
Koppången National Park	Dalarna County	5000 ha (unknown operator: u'strong' acres)	TBA
Reivo National Park	Norrbotten County	11000 ha (unknown operator: u'strong' acres)	TBA
Rogen-Juttulslätten National Park	Dalarna County and Jämtland County	100000 ha (unknown operator: u'strong' acres)	TBA
Sankt Anna National Park	Östergötland County	10000 ha (unknown operator: u'strong' acres)	TBA

Vindelfjällen National Park	Västerbotten County	550000 ha (unknown operator: u'strong' acres)	TBA

See also

- List of World Heritage Sites in Sweden

References

[1] "National parks and other ways to protect nature" (http://www.naturvardsverket.se/en/In-English/Start/Nature-conservation-and-wildlife/ Nature-conservation-and-species-protection/National-parks-and-other-ways-to-protect-nature/). Naturvårdsverket (Swedish Environmental Protection Agency). . Retrieved 2009-06-11.

[2] "Nationalparker" (http://www.naturvardsverket.se/sv/Start/Naturvard/Skydd-av-natur/Nationalparker/) (in Swedish). Naturvårdsverket (Swedish Environmental Protection Agency). . Retrieved 2011-06-18.

[3] "Ny nationalparksplan för Sverige" (http://www.naturvardsverket.se/Start/Naturvard/Skydd-av-natur/Nationalparker/ Forslag-till-nya-nationalparker/Ny-nationalparksplan/) (in Swedish). Naturvårdsverket (Swedish Environmental Protection Agency). . Retrieved 2009-06-11.

[4] "Fakta om nationalparker" (http://www.naturvardsverket.se/Start/Naturvard/Skydd-av-natur/Nationalparker/ Forslag-till-nya-nationalparker/Fakta-om-nationalparker/) (in Swedish). Naturvårdsverket (Swedish Environmental Protection Agency). . Retrieved 2011-06-18.

[5] "Sarek National Park" (http://www.naturvardsverket.se/en/In-English/Start/Enjoying-nature/ National-parks-and-other-places-worth-visiting/National-Parks-in-Sweden/Sarek-National-Park/). Naturvårdsverket (Swedish Environmental Protection Agency). . Retrieved 2009-02-26.

[6] "Padjelanta National Park" (http://www.naturvardsverket.se/en/In-English/Start/Enjoying-nature/ National-parks-and-other-places-worth-visiting/National-Parks-in-Sweden/Padjelanta-National-Park/). Naturvårdsverket (Swedish Environmental Protection Agency). . Retrieved 2009-02-26.

[7] "Laponian Area" (http://whc.unesco.org/pg.cfm?cid=31&id_site=774). UNESCO World Heritage Centre. . Retrieved 2009-06-11.

[8] "Söderåsen National Park" (http://www.naturvardsverket.se/en/In-English/Start/Enjoying-nature/ National-parks-and-other-places-worth-visiting/National-Parks-in-Sweden/Soderasen-National-Park/). Naturvårdsverket (Swedish Environmental Protection Agency). . Retrieved 2009-02-26.

[9] "Dalby Söderskog National Park" (http://www.naturvardsverket.se/en/In-English/Start/Enjoying-nature/ National-parks-and-other-places-worth-visiting/National-Parks-in-Sweden/Dalby-Soderskog-National-Park/). Naturvårdsverket (Swedish Environmental Protection Agency). . Retrieved 2009-02-26.

[10] "Stenshuvud National Park" (http://www.naturvardsverket.se/en/In-English/Start/Enjoying-nature/ National-parks-and-other-places-worth-visiting/National-Parks-in-Sweden/Stenshuvud-National-Park/). Naturvårdsverket Swedish Environmental Protection Agency. . Retrieved 2009-02-26.

[11] "Fulufjället National Park" (http://www.naturvardsverket.se/en/In-English/Start/Enjoying-nature/ National-parks-and-other-places-worth-visiting/National-Parks-in-Sweden/Fulufjallet-National-Park/). Naturvårdsverket (Swedish Environmental Protection Agency). . Retrieved 2009-02-26.

[12] "Vision" (http://web.archive.org/web/20090621213915/http://www.panparks.org/Introduction/Vision). PAN Parks. Archived from the original (http://www.panparks.org/Introduction/Vision) on 2009-06-21. . Retrieved 2009-06-25.

[13] "National Parks in Sweden" (http://www.naturvardsverket.se/en/In-English/Start/Enjoying-nature/ National-parks-and-other-places-worth-visiting/National-Parks-in-Sweden/). Naturvårdsverket (Swedish Environmental Protection Agency). . Retrieved 2009-06-11.

[14] "Abisko National Park" (http://www.naturvardsverket.se/en/In-English/Start/Enjoying-nature/ National-parks-and-other-places-worth-visiting/National-Parks-in-Sweden/Abisko-National-Park/). Naturvårdsverket (Swedish Environmental Protection Agency). . Retrieved 2009-02-26.

[15] "Ängsö National Park" (http://www.naturvardsverket.se/en/In-English/Start/Enjoying-nature/ National-parks-and-other-places-worth-visiting/National-Parks-in-Sweden/Angso-National-Park/). Naturvårdsverket (Swedish Environmental Protection Agency). . Retrieved 2009-02-26.

[16] "Björnlandet National Park" (http://www.naturvardsverket.se/en/In-English/Start/Enjoying-nature/ National-parks-and-other-places-worth-visiting/National-Parks-in-Sweden/Bjornlandet-National-Park/). Naturvårdsverket (Swedish Environmental Protection Agency). . Retrieved 2009-02-26.

[17] "Blå Jungfrun National Park" (http://www.naturvardsverket.se/en/In-English/Start/Enjoying-nature/ National-parks-and-other-places-worth-visiting/National-Parks-in-Sweden/Bla-Jungfrun-National-Park/). Naturvårdsverket (Swedish Environmental Protection Agency). . Retrieved 2009-02-26.

[18] "Djurö National Park" (http://www.naturvardsverket.se/en/In-English/Start/Enjoying-nature/ National-parks-and-other-places-worth-visiting/National-Parks-in-Sweden/Djuro-National-Park/). Naturvårdsverket (Swedish

Environmental Protection Agency). . Retrieved 2009-02-26.

[19] "Färnebofjärden National Park" (http://www.naturvardsverket.se/en/In-English/Start/Enjoying-nature/
      National-parks-and-other-places-worth-visiting/National-Parks-in-Sweden/Farnebofjarden-National-Park/). Naturvårdsverket (Swedish
      Environmental Protection Agency). . Retrieved 2009-02-26.

[20] "Garphyttan National Park" (http://www.naturvardsverket.se/en/In-English/Start/Enjoying-nature/
      National-parks-and-other-places-worth-visiting/National-Parks-in-Sweden/Garphyttan-National-Park/). Naturvårdsverket (Swedish
      Environmental Protection Agency). . Retrieved 2009-02-26.

[21] "Gotska Sandön National Park" (http://www.naturvardsverket.se/en/In-English/Start/Enjoying-nature/
      National-parks-and-other-places-worth-visiting/National-Parks-in-Sweden/Gotska-Sandon-National-Park/). Naturvårdsverket (Swedish
      Environmental Protection Agency). . Retrieved 2009-02-26.

[22] "Hamra National Park" (http://www.naturvardsverket.se/en/In-English/Start/Enjoying-nature/
      National-parks-and-other-places-worth-visiting/National-Parks-in-Sweden/Hamra-National-Park/). Naturvårdsverket (Swedish
      Environmental Protection Agency). . Retrieved 2009-02-26.

[23] "Haparanda Archipelago National Park" (http://www.naturvardsverket.se/en/In-English/Start/Enjoying-nature/
      National-parks-and-other-places-worth-visiting/National-Parks-in-Sweden/Haparanda-Skargard-National-Park/). Naturvårdsverket (Swedish
      Environmental Protection Agency). . Retrieved 2009-02-26.

[24] "Kosterhavets webbplats [Website of the Kosterhavet Marine National Park]" (http://www.lansstyrelsen.se/vastragotaland/
      Projektwebbar/Kosterhavet/) (in Swedish). Västra Götaland County Administrative Board. . Retrieved 2009-09-10.

[25] "Kosterhavet blir Sveriges första marina nationalpark" (http://web.archive.org/web/20090330195818/http://www.naturvardsverket.se/
      sv/Arbete-med-naturvard/Satsning-pa-havsmiljo/Skydd-av-marina-omraden/Kosterhavet-planeras-bli-Sveriges-forsta-marina-nationalpark/
) (in Swedish). Naturvårdsverket (Swedish Environmental Protection Agency). Archived from the original (http://www.naturvardsverket.se/
      sv/Arbete-med-naturvard/Satsning-pa-havsmiljo/Skydd-av-marina-omraden/Kosterhavet-planeras-bli-Sveriges-forsta-marina-nationalpark/
) on 2009-03-30. . Retrieved 2009-06-26.

[26] "Sveriges första marina nationalpark" (http://www.svd.se/nyheter/inrikes/artikel_2618793.svd) (in Swedish). Svenska Dagbladet.
      2009-03-19. . Retrieved 2009-09-10.

[27] "Kosterhavet National Park" (http://www.naturvardsverket.se/en/In-English/Start/Enjoying-nature/
      National-parks-and-other-places-worth-visiting/National-Parks-in-Sweden/Kosterhavet-National-Park/). Naturvårdsverket (Swedish
      Environmental Protection Agency). . Retrieved 2009-09-14.

[28] "Muddus National Park" (http://www.naturvardsverket.se/en/In-English/Start/Enjoying-nature/
      National-parks-and-other-places-worth-visiting/National-Parks-in-Sweden/Muddus-National-Park/). Naturvårdsverket (Swedish
      Environmental Protection Agency). . Retrieved 2009-02-26.

[29] "Norra Kvill National Park" (http://www.naturvardsverket.se/en/In-English/Start/Enjoying-nature/
      National-parks-and-other-places-worth-visiting/National-Parks-in-Sweden/Norra-Kvill-National-Park/). Naturvårdsverket (Swedish
      Environmental Protection Agency). . Retrieved 2009-02-26.

[30] "Pieljekaise National Park" (http://www.naturvardsverket.se/en/In-English/Start/Enjoying-nature/
      National-parks-and-other-places-worth-visiting/National-Parks-in-Sweden/Pieljekaise-National-Park/). Naturvårdsverket (Swedish
      Environmental Protection Agency). . Retrieved 2009-02-26.

[31] "Sånfjället National Park" (http://www.naturvardsverket.se/en/In-English/Start/Enjoying-nature/
      National-parks-and-other-places-worth-visiting/National-Parks-in-Sweden/Sanfjallet-National-Park/). Naturvårdsverket (Swedish
      Environmental Protection Agency). . Retrieved 2009-02-26.

[32] "Skuleskogen National Park" (http://www.naturvardsverket.se/en/In-English/Start/Enjoying-nature/
      National-parks-and-other-places-worth-visiting/National-Parks-in-Sweden/Skuleskogen-National-Park/). Naturvårdsverket (Swedish
      Environmental Protection Agency). . Retrieved 2009-02-26.

[33] "Stora Sjöfallet National Park" (http://www.naturvardsverket.se/en/In-English/Start/Enjoying-nature/
      National-parks-and-other-places-worth-visiting/National-Parks-in-Sweden/Stora-Sjofallet-National-Park/). Naturvårdsverket (Swedish
      Environmental Protection Agency). . Retrieved 2009-02-26.

[34] "Store Mosse National Park" (http://www.naturvardsverket.se/en/In-English/Start/Enjoying-nature/
      National-parks-and-other-places-worth-visiting/National-Parks-in-Sweden/Store-Mosse-National-Park/). Naturvårdsverket (Swedish
      Environmental Protection Agency). . Retrieved 2009-02-26.

[35] "Tiveden National Park" (http://www.naturvardsverket.se/en/In-English/Start/Enjoying-nature/
      National-parks-and-other-places-worth-visiting/National-Parks-in-Sweden/Tiveden-National-Park/). Naturvårdsverket (Swedish
      Environmental Protection Agency). . Retrieved 2009-02-26.

[36] "Töfsingdalen National Park" (http://www.naturvardsverket.se/en/In-English/Start/Enjoying-nature/
      National-parks-and-other-places-worth-visiting/National-Parks-in-Sweden/Tofsingdalen-National-Park/). Naturvårdsverket (Swedish
      Environmental Protection Agency). . Retrieved 2009-02-26.

[37] "Tresticklan National Park" (http://www.naturvardsverket.se/en/In-English/Start/Enjoying-nature/
      National-parks-and-other-places-worth-visiting/National-Parks-in-Sweden/Tresticklan-National-Park/). Naturvårdsverket (Swedish
      Environmental Protection Agency). . Retrieved 2009-02-26.

[38] "Tyresta National Park" (http://www.naturvardsverket.se/en/In-English/Start/Enjoying-nature/
      National-parks-and-other-places-worth-visiting/National-Parks-in-Sweden/Tyresta-National-Park/). Naturvårdsverket (Swedish
      Environmental Protection Agency). . Retrieved 2009-02-26.

[39] "Vadvetjåkka National Park" (http://www.naturvardsverket.se/en/In-English/Start/Enjoying-nature/
      National-parks-and-other-places-worth-visiting/National-Parks-in-Sweden/Vadvetjakka-National-Park/). Naturvårdsverket (Swedish
      Environmental Protection Agency). . Retrieved 2009-02-26.

[40] "Förslag till nya nationalparker" (http://web.archive.org/web/20081225000717/http://www.naturvardsverket.se/sv/
      Arbete-med-naturvard/Skydd-och-skotsel-av-vardefull-natur/Nationalparker/Forslag-till-nya-nationalparker/) (in Swedish).
      Naturvårdsverket (Swedish Environmental Protection Agency). Archived from the original (http://www.naturvardsverket.se/sv/
      Arbete-med-naturvard/Skydd-och-skotsel-av-vardefull-natur/Nationalparker/Forslag-till-nya-nationalparker/) on 2008-12-25. . Retrieved
      2009-06-11.

Dalby,_Lund

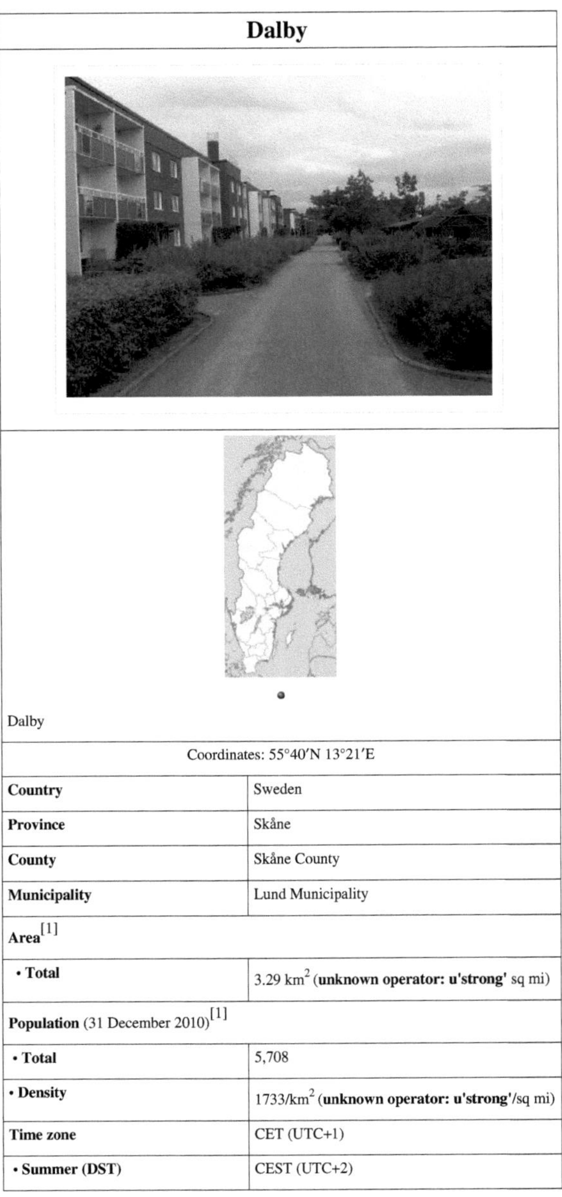

<table>
<tr><th colspan="2" align="center">Dalby</th></tr>
<tr><td colspan="2" align="center">Dalby</td></tr>
<tr><td colspan="2" align="center">Coordinates: 55°40′N 13°21′E</td></tr>
<tr><td>Country</td><td>Sweden</td></tr>
<tr><td>Province</td><td>Skåne</td></tr>
<tr><td>County</td><td>Skåne County</td></tr>
<tr><td>Municipality</td><td>Lund Municipality</td></tr>
<tr><td colspan="2">Area[1]</td></tr>
<tr><td>• Total</td><td>3.29 km^2 (unknown operator: u'strong' sq mi)</td></tr>
<tr><td colspan="2">Population (31 December 2010)[1]</td></tr>
<tr><td>• Total</td><td>5,708</td></tr>
<tr><td>• Density</td><td>1733/km^2 (unknown operator: u'strong'/sq mi)</td></tr>
<tr><td>Time zone</td><td>CET (UTC+1)</td></tr>
<tr><td>• Summer (DST)</td><td>CEST (UTC+2)</td></tr>
</table>

Dalby is a locality situated in Lund Municipality, Skåne County, Sweden with 5,708 inhabitants in 2010.[1] It is located about 10 km east-south-east of Lund, and about 20 km east-north-east of Malmö.

Dalby has the oldest stone church in Scandinavia, to which the cathedral in Hildesheim served as a model. In 1060 the Danish King Svend Estridsen initiated the creation of a religious centre in Dalby, and also constructed his King's Residence here. From 1060 Dalby was a bishopric under the bishop Egino, appointed by Adalbert of Bremen. But as early as 1066 the English antibishop Henrik was enthroned in Lund, probably elected by the people and the clergy. In 1085 Canute the Saint decided to build a new cathedral in Lund and in 1104 Lund became the archbishopric over Scandinavia.

Until the Protestant Reformation in Denmark in 1536, Dalby retained some importance as the site of an Augustinian monastery and a demesne of the Crown.

Dalby was a municipality up until 1974, when it became part of Lund municipality. Between 1941 and 1954, Dalby was also a municipal urban area (*"municipalsamhälle"*).

The Dalby Söderskog national park is situated just northwest of Dalby.

The old quarry *Stenbrottet* is located a couple of kilometers east of Dalby. It's a popular place for swimming and fishing and for couples to become engaged.

The Holy Cross Church in Dalby

References

[1] "Tätorternas landareal, folkmängd och invånare per km^2 2005 och 2010" (http://www.scb.se/ Statistik/MI/MI0810/2010A01/ Tatorternami0810tab1_4.xls) (in Swedish). Statistics Sweden. 14 December 2011. Archived (http://www.webcitation.org/64arqC15e) from the original on 10 January 2012. . Retrieved 10 January 2012.

Viborg_Municipality

Viborg municipality is a municipality (Danish, *kommune*) in Region Midtjylland on the Jutland peninsula in northern Denmark. The municipality covers an area of 1,390 km², and has a total population of 92,084 (2008). Its mayor is Johannes Stensgaard, a member of the Social Democrats (*Socialdemokraterne*) political party.

The main town and the site of its municipal council is the town of Viborg.

On January 1, 2007 Viborg municipality was, as the result of *Kommunalreformen* ("The Municipal Reform" of 2007), merged with Bjerringbro, Fjends, Karup, Møldrup, and Tjele municipalities to form an enlarged Viborg municipality.

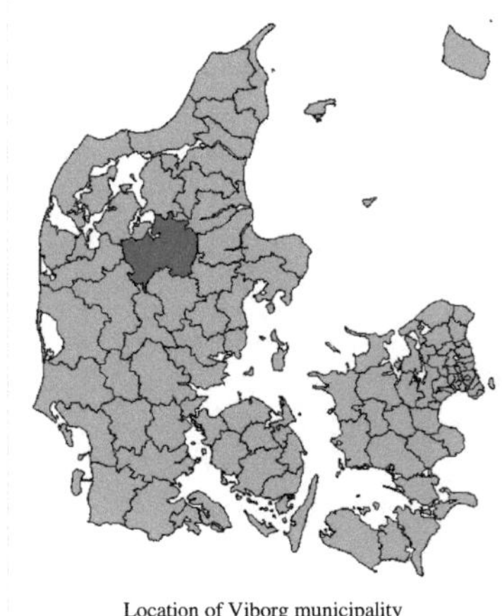

Location of Viborg municipality

Twin cities

Viborg has a twin city in each of the Nordic countries, as well as in other world regions.

- Porvoo, Finland
- Dalvík, Iceland
- Greifswald, Germany
- Hamar, Norway
- Kecskemét, Hungary
- León, Nicaragua
- Lund Municipality, Sweden
- Nevers, France
- Zabrze, Poland

See also

- World Map at Lake Klejtrup

Cathedral in Viborg, Denmark. ©2001 Hans Andersen.

External links

- Municipality's official website [1]

References

- Municipal statistics: NetBorger Kommunefakta [2], delivered from KMD aka Kommunedata (Municipal Data) [3]
- Municipal mergers and neighbors: Eniro new municipalities map [4]
- Searchable/printable municipal maps: Krak mapsearch [5](outline visible but doesn't print out!)

References

[1] http://www.viborg.dk
[2] http://www2.netborger.dk/Kommunefakta/
[3] http://www.kmd.dk/
[4] http://kommune.eniro.dk/danmarkskort/
[5] http://kort.krak.dk/borgerdk.kortsoegning/imapDKbig.asp

Hamar

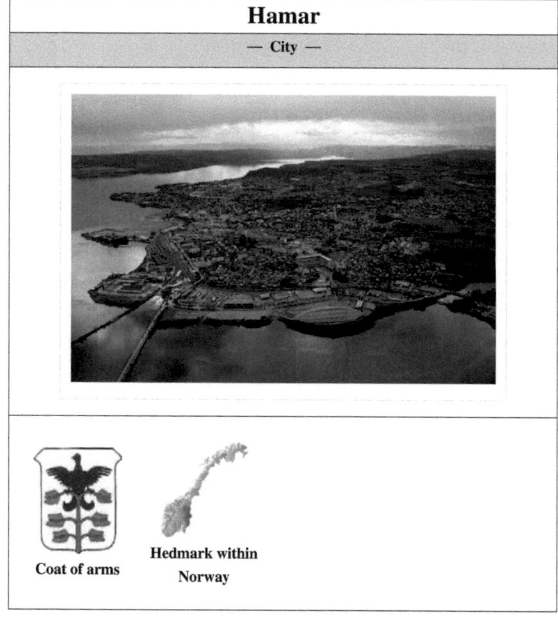

Hamar within Hedmark

Coordinates: 60°47′57″N 11°3′22″E

Country	Norway
County	Hedmark
District	Hedmarken
Municipality ID	NO-0403
Administrative centre	Hamar
Government	
• **Mayor (2012)**	Morten Aspeli (Labour Party)
Area(Nr. 257 in Norway)	
• **Total**	351 km^2 (**unknown operator: u'strong'** sq mi)
• **Land**	338 km^2 (**unknown operator: u'strong'** sq mi)
Population (2006)	
• **Total**	28,996
• **Change (10 years)**	4.3 %
• **Rank in Norway**	31
Time zone	CET (UTC+1)
• **Summer (DST)**	CEST (UTC+2)
Official language form	Neutral
Demonym	Hamarenser/Hamarensar Hamarsing [1]
Website	www.hamar.kommune.no [2]
Data from Statistics Norway [3]	

Hamar kommune	
— Municipality —	

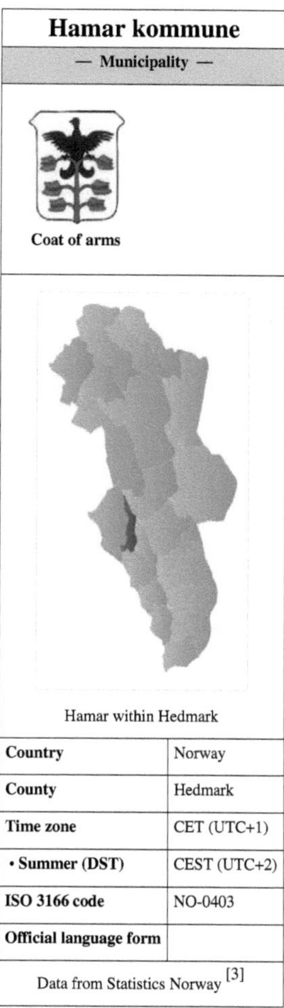

Coat of arms

Hamar within Hedmark

Country	Norway
County	Hedmark
Time zone	CET (UTC+1)
• Summer (DST)	CEST (UTC+2)
ISO 3166 code	NO-0403
Official language form	
Data from Statistics Norway [3]	

Hamar is a town and municipality in Hedmark county, Norway. It is part of the traditional region of Hedmarken. The administrative centre of the municipality is the town of Hamar. The municipality of Hamar was separated from Vang as a town and municipality of its own in 1849. Vang was merged back into Hamar on 1 January 1992.

The town is located on the shores of lake Mjøsa, Norway's largest lake, and is the principal city of Hedmark county. It is bordered to the northwest by the municipality of Ringsaker, to the north by Åmot, to the east by Løten, and to the south by Stange.

General information

Name

The municipality (originally the town) is named after the old *Hamar* farm (Old Norse: *Hamarr*), since the medieval town was built on its ground. The name is identical with the word *hamarr* which means "rocky hill".

Coat-of-arms

The coat-of-arms shows a Black Grouse sitting in the top of a pine tree on a white background. It was first described in the anonymous Hamarkrøniken (*The Hamar Chronicle*) written in 1553.

History

Between 500 and 1000 AD, Aker farm was one of the most important power centres in Norway, located just a few kilometres away from today's Hamar. Three coins found in Ringerike in 1895 have been dated to the time of Harald Hardråde and are inscribed *Olafr a Hamri*.

Middle Ages

At some point, presumably after 1030 but clearly before 1152, the centre was moved from Aker to the peninsula near Rosenlundvika, what we today know as Domkirkeodden. There are some indications Harald Hardråde initiated this move because he had property at the new site.

Much of the information about medieval Hamar is derived from the Hamar Chronicles, dated to about 1550. The town is said to have reached its apex in the early 14th century, dominated by the Hamar cathedral, bishop's manor, and fortress, and surrounding urbanization. The town was known for its fragrant apple orchards, but there were also merchants, craftsmen, and fishermen in the town.

After the Christianization of Norway in 1030, Hamar began to gain influence as a centre for trade and religion, until the episcopal representative Nikolaus Breakspear in 1152 founded Hamar Kaupangen as one of five dioceses in medieval Norway. This diocese included Hedemarken and Christians Amt, being separated in 1152 from the former diocese of Oslo. The first bishop of Hamar was Arnold, Bishop of Gardar, Greenland (1124–1152). He began to build the now ruined cathedral of Christ Church, which was completed about the time of Bishop Paul (1232–1252). Bishop Thorfinn (1278–1282) was exiled and died at Ter Doest abbey in Flanders. Bishop Jörund (1285–1286) was transferred to Trondheim. A provincial council was held in 1380. Hamar remained an important religious and political centre in Norway, organized around the cathedral and the bishop's manor until the Reformation in 1536, when it lost its status as a bishopric after the last Catholic bishop, Mogens Lauritsson (1513–1537), was taken prisoner in his castle at Hamar by Truid Ulfstand, a Danish noble, and sent to Antvorskov in Denmark, where he was mildly treated until his death in 1542. There were at Hamar a cathedral chapter with ten canons, a school, a Dominican Priory of St. Olaf, and a monastery of the Canons Regular of St. Anthony of Vienne.[4]

Hamar, like most of Norway, was severely diminished by the Black Plague in 1349, and by all accounts continued this decline until the Reformation, after which it disappeared.

The Reformation in Norway took less than 10 years to complete, from 1526 to 1536. The fortress was made into the residence of the sheriff and renamed Hamarhus fortress. The cathedral was still used but fell into disrepair culminating with the Swedish army's siege and attempted demolition in 1567, during the Northern Seven Years' War, when the manor was also devastated.

Reformation and decline

By 1587, merchants in Oslo had succeeded in moving all of Hamar's market activities to Oslo. Though some regional and seasonal trade persisted into the 17th century, Hamar as a town ceased to exist by then. In its place, the area was used for agriculture under the farm of Storhamar, though the ruins of the cathedral, fortress, and lesser buildings became landmarks for centuries since then.

The King made Hamarhus a feudal seat until 1649, when Frederick III transferred the property known as Hammer to Hannibal Sehested, making it private property. In 1716, the estate was sold to Jens Grønbech (1666–1734). With this, a series of construction projects started, and the farm became known as Storhamar, passing through several owners until Norwegian nobility was abolished in 1831, when Erik Anker took over the farm.[5]

The founding of modern Hamar

As early as 1755, the Danish government in Copenhagen expressed an interest in establishing a trading center on Mjøsa. Elverum was considered a frontier town with frequent unrest, and there was even talk of encouraging the dissenting Hans Nielsen Hauge to settle in the area. Bishop Fredrik Julius Bech, one of the most prominent officials of his time, proposed establishing a town at or near Storhamar, at the foot of Furuberget.

In 1812, negotiations started in earnest, when the regional governor of Kristians Amt, proposed establishing a market on Mjøsa. A four-person commission was named on 26 July 1814, with the mandate of determining a suitable site for a new town along the shore. On 8 June 1815, the commission recommended establishing such a town at Lillehammer, then also a farm, part of Fåberg.

Acting on objections to this recommendation, the department of the interior asked two professors, Ludvig Stoud Platou and Gregers Fougner Lundh, to survey the area and develop an alternative recommendation. It appears that Lundh in particular put great effort into this assignment, and in 1824 he presented to the Storting a lengthy report, that included maps and plans for the new town.

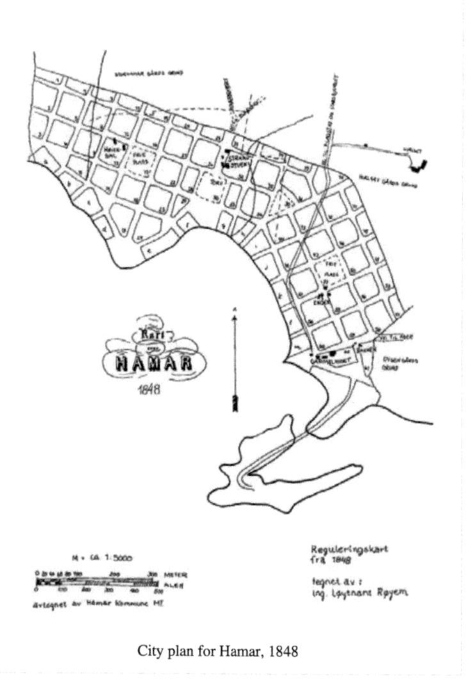

City plan for Hamar, 1848

Lundh's premise was that the national economic interest reigned supreme, so he based his recommendation on the proposed town's ability to quickly achieve self-sustaining growth. He proposed that the name of the new town be called *Carlshammer* and proposed it be built along the shore just north of Storhamar and eastward. His plans were detailed, calling for streets 20 meters broad, rectangular blocks with 12 buildings in each, 2 meters separating each of them. He also proposed tax relief for 20 years for the town's first residents, that the state relinquish property taxes in favor of the city, and that the city be given monopoly rights to certain trade. He even proposed that certain types of foreigners be allowed to settle in the town to promote trade, in particular, the Quakers.

His recommendation was accepted in principle by the government, but the parliamentary committee equivocated on the location. It left the determination of the actual site to the king so as to not slow down things further. Another commission was named in June 1825, consisting of Herman Wedel-Jarlsberg, professor Lundh, and other prominent Norwegians. After surveying the entire lake, it submitted another report that considered eleven different locations, including sites near today's Eidsvoll, Minnesund, Tangen in Stange, Aker, Storhamar, Brumunddal, Nes, Moelven, Lillehammer, Gjøvik, and Toten. Each was presented with pros and cons. The commission itself was split between Lillehammer and Storhamar. The parliament finally decided on Lillehammer, relegating Hamar once more, it seemed, to be a sleepy agricultural area.

As steamboats were introduced on the lake, the urban elite developed an interest in the medieval Hamar, and in 1841, editorials appeared advocating the reestablishment of a town at Storhamar. By then the limitations of Lillehammer's location had also become apparent, in particular those of its shallow harbor. After a few more years of discussions and negotiations both regionally and nationally, member of parliament Frederik Stang put on the table once more the possibility of a town in or near Storhamar. The governor at the time, Frederik Hartvig Johan Heidmann, presented a thorough deliberation of possible specific locations, and ended up proposing the current site, at Gammelhusbukten.

On 26 April 1848, the king signed into law the establishment of Hamar on the grounds of the farms of Storhamar and Holset, along the shores of Mjøsa. The law stated that the town will be founded on the date its borders are settled, which turned out to be 21 March 1849, known as the merchant town of Hamar, with a trading zone within five kilometers of its borders.[5]

Building a city

The area of the new town covered 400 mål which is the equivalent to today's 40 hectares (**unknown operator: u'strong'** acres) (40 hectare). An army engineer, Røyem, drafted the initial plan. There would be three thoroughfares, at Strandgata, Torggata, and Grønnegate (the latter the name of a medieval road) and a grid system of streets between them. The orientation of the town was toward the shore. Røyem set aside space for three parks and a public square, and also room for a church just outside the town's borders.

View of Hamar in the 1890s

There were critics of the plan, pointing out that the terrain was hilly and not suitable for the proposed rigid grid. Some adjustments were made, but the plan was largely accepted and is evident in today's Hamar. There were also lingering concerns about the town's vulnerability to flooding.

No sooner had the ink dried on the new law, and building started in the spring of 1849. The first buildings were much like sheds, but there was great enthusiasm, and by the end of 1849, ten buildings were insured in the new town. None of these are standing today; the last two were adjacent buildings on Skappelsgate. By 1850, there were 31 insured houses, and 1852, 42; and in 1853, 56. Building slowed down for a few years and then picked up again in 1858, and by the end of 1860 there were 100 insured houses in the town. The shore side properties were obliged to grow gardens, setting the stage for a leafy urban landscape.

Roads quickly became a challenge – in some places, it was necessary to ford creeks in the middle of town. The road inspector found himself under considerable stress, and it was not until 1869 street names were settled. Highways in and out of the city also caused considerable debate, especially when it came to financing their construction.

The first passenger terminal in Hamar was in fact a crag in the lake, from which travelers were rowed into the city. In 1850, another pier was built with a two-storey terminal building. All this was complicated by the significant seasonal variations in water levels. In 1857 a canal was built around a basin that would allow freight ships to access a large warehouse. Although the canal and basin still were not deep enough to accommodate passenger steamships, the area became one of the busiest areas in the town and the point around which the harbor was further developed.[6]

The Diocese of Hamar was established in 1864, and the Hamar Cathedral was consecrated in 1866 and remains a central point in the city.

A promenade came into being from the harbor area, past the gardens on the shore, and north toward the site of the old town.

Establishment of government

The first executive of Hamar was Johannes Bay, who arrived in October 1849 to facilitate an election of a board of supervisors and representatives. The town's Royal Charter called for the election of 3 supervisors and 9 representatives, and elections were announced in the paper and through town crier. Of the 10 eligible town citizens, three supervisors were elected, and the remaining six were elected by consent to be representatives, resulting in a shortfall of 3 on the board. The first mayor of Hamar was Christian Borchgrevink.

The first order of business was the allocation of liquor licenses and the upper limit of alcohol that could be sold within the town limits. The board quickly decided to award licenses to both applicants and set the upper limit to 12,000 "pots" of liquor, an amount that was for all intents and purposes limitless.

The electorate increased in 1849 to 26, including merchants and various craftsmen, and the empty representative posts were filled in November. In 1850, the board allowed for unlimited exercise of any craft for which no citizenship had been taken out, which led to much unregulated craftsmanship. Part time policemen were hired, and the town started setting taxes and a budget by the end of 1849. In 1850, a new election was held for the town board.

The painter Jakobsen had early on offered his house for public meetings and assembly, and upon buying a set of solid locks, his basement also became the town prison. One merchant was designated as the town's firefighter and was given two buckets with equipment, and later a simple hose, but by 1852 a full time fire chief was named. There was also some controversy around the watchman who loudly reported the time to all the town's inhabitants every half hour, every night. Hamar also had a scrupulously enforced ordnance against smoking (pipe) without a lid in public or private.

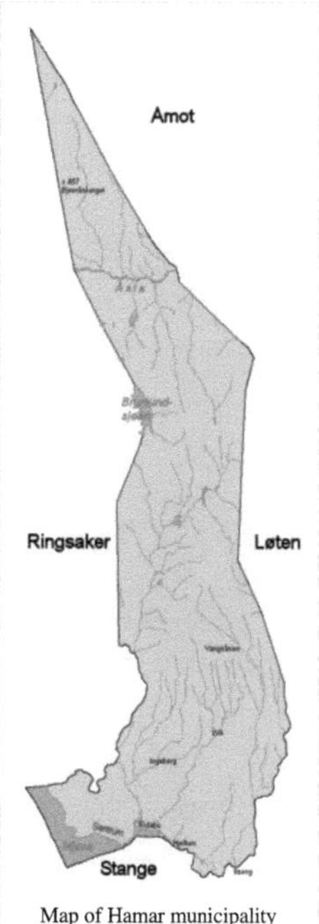

Map of Hamar municipality

In Hamar's early days, the entire population consisted of young entrepreneurs, and little was needed in the way of social services. After a few years, a small number of indigent people needed support, and a poorhouse was erected.[6]

Fires, floods and other disasters

In 1878, as the firefighting capabilities of the young town were upgraded, a fire broke out in a bakery that was put out without doing too much damage. In February 1879 at 2:00 in the morning another fire broke out after festivities, burning down an entire building that housed many historical items from town's history. This was followed by a series of fires that left entire blocks in ashes that seemed to come to an end in 1881, when a professional fire corps was hired.

In 1860, concerns about flooding were vindicated when a late and sudden spring caused the lake to flood, peaking on about 24 June, when the street-level floor of the front properties was completely inundated. This was the worst flood recorded since 1789. By 9 July, the floods had receded. But it was not to prove the end of the calamities. In August, massive rainfall led to flash flooding in the area, putting several streets under water. This was immediately followed by unseasonably cold weather, freezing the potato crops and inconveniencing Hamar's residents. And then, mild weather melted all the ice and accumulated snow, leading to another round of flooding. By the time a particularly cold and snow-filled winter set in, there was mostly relief about getting some stability.

In 1876, the town was scandalized by the apprehension of one Kristoffer Svartbækken, arrested for the cold-blooded murder of 19-year old Even Nilsen Dæhlin. Svartbækken was convicted for the murder and executed the year after in the neighboring rural community of Løten in what must have been a spectacle with an audience of 3,000 locals, presumably most of Hamar's population at the time.

Then in 1889, there were riots in Hamar over the arrest of one of their own constables, one sergeant Huse, who had been insubordinate while on a military drill at the cavalry camp at Gardermoen. In an act of poor judgment, Huse's superior sent him to Hamar's prison in place of military stockades. Partly led and partly tolerated by other constables, the town's population engaged in demonstrations, marches, and other unlawful but non-violent acts that were effectively ended when a company of soldiers arrived from the camp at Terningmoen near Elverum.[6]

Hamar panorama

Cityscape

The Hedmark museum, located on Domkirkeodden, is an important historical landmark in Hamar, an outdoor museum with remains of the medieval church, in a protective glass housing, the episcopal fortress and a collection of old farm houses. The museum is a combined medieval, ethnological and archaeological museum and has received architectural prizes for its approach to conservation and exhibition. It also houses a vast photographic archive for the Hedmark region.

Hamar is also known for its indoor long track speed skating and bandy arena, the *Olympia Hall*, better known as Vikingskipet ("The Viking ship") for its shape. It was built to host the speed skating competitions of the 1994 Winter Olympics that were held in nearby Lillehammer. Already in 1993 it hosted the Bandy World Championships.The Vikingskipet Olympic Arena was later used in the winter of 2007 as the service park for Rally Norway, the second round of the 2007 World Rally Championship season. Its been the host for the worlds second largest computer party The Gathering starting on the Wednesday in Easter each year, for the last 13 years.

Also situated in Hamar is the Hamar Olympic Amphitheatre which hosted the figure skating and short track speed skating events of the 1994 Winter Olympics. The figure skating competition was highly anticipated. It featured Nancy Kerrigan and Tonya Harding, who drew most of the media attention, however the gold medal was won by Oksana Baiul of Ukraine.

The centre of Hamar is the pedestrian walkway in the middle of town, with the library, cinema and farmer's market on Stortorget (the big square) on the western side, and Østre Torg (the eastern square), which sits on top of an underground multi-story carpark, on the eastern side.

Transport

Hamar is an important railway junction between two different lines to Trondheim. Rørosbanen, the old railway line, branches off from the mainline Dovrebanen. The Norwegian national railway museum (*Norsk Jernbanemuseum*) is also situated in Hamar.

Hamar's pedestrian street

The Viking Ship, a speed skating and bandy arena

Hamar Olympic Amphitheatre

Domkirkeruinene (cathedral ruins)

Ruins of the Hamar cathedral

building at the Hedmarks museum

Hamar Railway Station

Åsta river

Notable residents

Main category: People from Hamar

- Anders Baasmo Christiansen, actor (b. 1976)
- Rut Brandt, writer (1920 - 1986)
- Irene Dalby, former top swimmer (b. 1971)
- Egil Danielsen, Olympic gold medalist, javelin (b. 1933)
- Sigurd Evensmo, author (1912-1978)
- Knut Faldbakken, writer (b. 1941)
- Kirsten Flagstad, opera singer (1895-1962)
- Hulda Garborg, writer (1862-1934)
- Rolf Jacobsen, Poet (1907-1994)
- Olaf Johannessen, shooter (1890-1977)
- Erik Kristiansen, former ice hockey player (b. 1963)
- Terje Kojedal, former professional footballer (b. 1957)
- Torill Kove, award-winning animator (b. 1958)
- Katti Anker Møller, civil rights activist (1868-1945)
- Vegard Skogheim, former professional footballer (b. 1966)

- Patrick Thoresen, professional ice hockey player (b. 1983)
- Anette Trettebergstuen, politician (b. 1981)
- Petter Vaagan Moen, footballer (b. 1984)
- Even Wetten, former speed skater (b. 1982)
- Ole Edvard Antonsen, Musician trumpet player (b. 1962)

Sports

Team sports

Hamar boasts several teams at the Norwegian top level in various sports:

- Hamarkameratene (Ham-Kam) is a football (soccer) team that plays in the Adeccoligaen after being relegated from the Tippeligaen in 2008.
- Storhamar Dragons is an ice hockey team which is the reigning Norwegian champion after winning the playoffs in 2008. The club has won the title a total of six times. Storhamar has been dominating the GET-ligaen in Norway since the beginning of the 1990s.
- Storhamar Håndball is a handball team that has played one season in the Gildeserien, placing third in their first season.
- Fart IL is a women's football team currently playing its first season in the top league.
- Hamar Idrettslag has played in the highest bandy division recently, but this season, 2009–2010, they play in the 2nd.

Individual sports

Hamar is known for its speed skating history, both for its skaters and the championships that have been hosted by the city, already in 1894 Hamar hosted its first European championship, and the first World Championship the year after. After the Vikingskipet was built, Hamar has hosted international championships on a regular basis.

The most notable skaters from Hamar are Dag Fornæss and Even Wetten, both former World champions, allround and 1000m respectively. Amund Sjøbrend, Ådne Sønderål and Eskil Ervik have all been members of the local club Hamar IL, although they were not born in Hamar.

In Hamar on 17 July 1993, Scottish cyclist Graeme Obree set a world record for the distance covered in an hour. His 51,596 metres broke the 51,151 set at altitude nine years earlier but lasted only six days before Chris Boardman broke it in Bordeaux.

Other notable athletes:

- Egil Danielsen, javelin
- Irene Dalby, swimming
- Kamilla Gamme, diving
- Jan Frode Andersen, tennis
- Patrick Thoresen, ice hockey

International relations

Twin towns — Sister cities

The following cities, both in Scandinavia and around the world, are twinned with Hamar:[7]

- Lund, Skåne County, Sweden
- Viborg, Region Midtjylland, Denmark
- Dalvík, Iceland
- Porvoo, Uusimaa, Finland
- Greifswald, Mecklenburg-Vorpommern, Germany
- Khan Younis, Gaza strip, Palestinian Authority
- Fargo, North Dakota, United States
- Karmiel, North District, Israel

References

[1] "Personnemningar til stadnamn i Noreg" (http://www.sprakrad.no/Sprakhjelp/Rettskriving_Ordboeker/Innbyggjarnamn) (in Norwegian).
     Språkrådet. .
[2] http://www.hamar.kommune.no/
[3] http://www.ssb.no/english/municipalities/0403
[4] This article incorporates text from a publication now in the public domain: Herbermann, Charles, ed. (1913). "Ancient See of Hamar".
     Catholic Encyclopedia. Robert Appleton Company.
[5] Ramseth, Christian (1991). *Hamar bys historie: til 50 års jubilæet 21 mars 1899*. Hamar Historical Society. (Norwegian)
[6] Pedersen, Ragnar (1990). *Den adelige frie sedegård Storhamar (Fra Kaupang og Bygd)*. Hedmarksmuseet og Domkirkeodden.
     ISBN 8299075270. (Norwegian)
[7] "Vennskapsbyer" (http://www.hamar.kommune.no/hamar/www/site/kultur_og_fritid/vennskapsbyer/). Hamar kommune. . Retrieved
     2008-12-22. (Norwegian)

External links

- Municipal fact sheet (http://www.ssb.no/english/municipalities/0403) from Statistics Norway
- Hamar travel guide from Wikitravel
- Hamar Pictorial click-through (http://hamar.clickwalk.no/indexe.html)
- The Hedmark Museum (http://www.hedmarksmuseet.no/) (Norwegian) (Translate: Google (http://translate.
  google.com/translate?u=http://www.hedmarksmuseet.no/&hl=en&ie=UTF-8&sl=no&tl=en), Babelfish
  (http://babelfish.yahoo.com/translate_url?trurl=http://www.hedmarksmuseet.no/&lp=no_en))
- Kirsten Flagstad Museum (http://www.kirsten-flagstad.no/) (Norwegian) (Translate: Google (http://translate.
  google.com/translate?u=http://www.kirsten-flagstad.no/&hl=en&ie=UTF-8&sl=no&tl=en), Babelfish
  (http://babelfish.yahoo.com/translate_url?trurl=http://www.kirsten-flagstad.no/&lp=no_en))
- The Norwegian national railway museum (http://www.norsk-jernbanemuseum.no/) (Norwegian) (Translate:
  Google (http://translate.google.com/translate?u=http://www.norsk-jernbanemuseum.no/&hl=en&
  ie=UTF-8&sl=no&tl=en), Babelfish (http://babelfish.yahoo.com/translate_url?trurl=http://www.
  norsk-jernbanemuseum.no/&lp=no_en))
- The Viking Ship (http://www.hoa.no/) (Norwegian) (Translate: Google (http://translate.google.com/
  translate?u=http://www.hoa.no/&hl=en&ie=UTF-8&sl=no&tl=en), Babelfish (http://babelfish.yahoo.
  com/translate_url?trurl=http://www.hoa.no/&lp=no_en))

Porvoo

<table>
<tr><td colspan="2" align="center">Porvoo
Porvoo – Borgå</td></tr>
<tr><td colspan="2" align="center">— City —</td></tr>
<tr><td colspan="2" align="center">Porvoon kaupunki
Borgå stad</td></tr>
<tr><td colspan="2" align="center">
Riverside storage buildings in Old Porvoo</td></tr>
<tr><td colspan="2" align="center">
Coat of arms</td></tr>
<tr><td colspan="2" align="center"></td></tr>
<tr><td>Country</td><td>Finland</td></tr>
<tr><td>Region</td><td>Uusimaa</td></tr>
<tr><td>Sub-region</td><td>Porvoo sub-region</td></tr>
<tr><td>City rights</td><td>ca. 1380</td></tr>
<tr><td colspan="2">Government</td></tr>
<tr><td>• City manager</td><td>Jukka-Pekka Ujula</td></tr>
<tr><td colspan="2">Area(2011-01-01)[1]</td></tr>
</table>

• Total	2139.19 km² (**unknown operator: u'strong'** sq mi)
• Land	654.70 km² (**unknown operator: u'strong'** sq mi)
• Water	1484.49 km² (**unknown operator: u'strong'** sq mi)
Area rank	38 largest in Finland
Population (2012-01-31)[2]	
• Total	48862
• Rank	21 largest in Finland
• Density	74.63/km² (**unknown operator: u'strong'**/sq mi)
Population by native language[3]	
• Finnish	64.9% (official)
• Swedish	31.6% (official)
• Others	3.5%
Population by age[4]	
• 0 to 14	18.5%
• 15 to 64	67%
• 65 or older	14.5%
Time zone	EET (UTC+2)
• Summer (DST)	EEST (UTC+3)
Municipal tax rate[5]	19.25%
Website	www.porvoo.fi [6]

Porvoo (Finnish pronunciation: [ˈporʋoː]; Swedish: *Borgå* Swedish pronunciation: [ˈbɔrɡo]) is a city and a municipality situated on the southern coast of Finland approximately 50 kilometres (**unknown operator: u'strong'** mi) east of Helsinki. Porvoo is one of the six medieval towns in Finland, first mentioned as a city in texts from 14th century. Porvoo is the seat of the Swedish speaking Diocese of Borgå of the Evangelical Lutheran Church of Finland.

History

Porvoo was first mentioned in documents in the early 14th century, and Porvoo was given city rights around 1380, even though according to some sources the city was founded in 1346. The old city of Porvoo was formally disestablished and the new city of Porvoo founded in 1997 when the city of Porvoo and the Rural municipality of Porvoo were consolidated.[7] When Sweden lost the city of Viborg to Russia in 1721, the episcopal seat was moved to Porvoo. At this time, Porvoo was the second largest city in Finland. After the conquest of Finland by Russian armies in 1808 Sweden had to cede Finland to Russia in 1809 (the Treaty of Fredrikshamn). The Diet of Porvoo in 1809 was a landmark in the History of Finland. The Tsar Alexander I confirmed the new Finnish constitution (which was essentially the Swedish constitution from 1772), and made Finland an autonomous Grand Duchy.

The *Porvoo Common Statement* is a report issued at the conclusion of theological conversations by official representatives of four Anglican Churches and eight Nordic and Baltic Lutheran Churches in 1989–1992. It established the Porvoo Communion, so named after the Porvoo Cathedral where the Eucharist was celebrated on the final Sunday of the conversations leading to the Statement.

Name

The town received its name from a Swedish earth fortress near the river Porvoonjoki which flows through the town. The name *Porvoo* is the Fennicised version of the Swedish name (*Borgå*) and its parts of *borg* meaning "castle" and *å* "river".[8]

Politics

Results of the Finnish parliamentary election, 2011 in Porvoo:

- Swedish People's Party 25.2%
- Social Democratic Party 19.2%
- National Coalition Party 18.0%
- True Finns 16.5%
- Green League 7.7%
- Centre Party 5.3%
- Left Alliance 4.7%
- Christian Democrats 2.2%

Urban development

The town is famed for its "Old Town" (*Gamla Stan* in Swedish), a dense mediaeval street pattern with predominantly wooden houses. The Old Town came close to being demolished in the 19th century by a new urban plan for the city. The plan was cancelled due to a popular resistance headed by Count Louis Sparre

The central point of the old town is the medieval, stone and brick Porvoo Cathedral which gave its name to the Porvoo Communion – an inter-church agreement between a number of Anglican and Lutheran denominations. The cathedral was damaged by fire on 29 May 2006: the roof was totally destroyed but the interior is largely intact. A drunken youth had played with fire at the church, unaware of recent tarwork and nearby tarcontainers, accidentally causing a large fire. He was later sentenced to a short prison term and restitutions of 4.3 million Euro.[9]

The red-coloured wooden storage buildings on the riverside are a proposed UNESCO world heritage site. Already by the early 19th century the authorities understood the value of the old town, and so with the need for growth a plan was made for a 'new town' built adjacent to the old town, following a grid plan but with houses also built in wood.

New Housing, Porvoo, by architect Tuomas Siitonen

By the end of the 20th century there was pressure to develop the essentially untouched western side of the river. There was concern that growth would necessitate the construction of a second bridge across the river into the town, thus putting further strain on the wooden town. An architectral competition was held in 1990, the winning entry of which proposed building the second bridge. Plans for the western side of the river have progressed under the direction of architect Tuomas Siitonen, and both a vehicle bridge and a pedestrian bridge have been built. The design for new housing is based on a typology derived from the old store houses on the opposite side of the river. Yet another new development entails the construction of a large business park called King's Gate (*Kuninkaanportti* in Finnish, *Kungsporten* in Swedish), which is presently under construction.

Porvoo railway station does not receive regular train services, but special excursion trains from Kerava (either with steam locomotives or comprising former VR diesel railcars from the 1950s) operate on summer Saturdays.[10]

Suomenkylä

Suomenkylä, or *Finnby* in Swedish, is a village north of the center of Porvoo and beside the Porvoo river. Suomenkylä has an old school founded by Johannes Linnankoski in 1898. The village of Suomenkylä also has two burial places from Bronze Age.

Kerkkoo

Kerkkoo, or *Kerko* in Swedish, is a village north of the center of Porvoo and beside the Porvoo river. Kerkkoo has an old school which is over 100 years old and still active. From the village of Kerkkoo archeologists and townspeople found a stone axe from the Bronze Age.

Sports

The local team Porvoon Akilles, or just Akilles, plays in the highest bandy division and has become Finnish champions twice. Sami Hyypiä,a football player for the Finnish national team is the main sports pride of Porvoo.

Notable people

- Remu Aaltonen, musician
- Torvald Appelroth, Olympic fencer
- Albert Edelfelt, painter
- Hanna Ek, Miss Finland 2005
- Sami Hyypiä, soccer player
- Johan Ludvig Runeberg, national poet
- Seppo Telenius, writer and historian

Porvoo Cathedral prior to the fire in May 2006

- Ville Vallgren, sculptor
- Vladimir Kirillovich, Grand Duke of Russia, Romanov Dynast

International relations

Twin towns — Sister cities

Porvoo is twinned with the following cities:[11] :

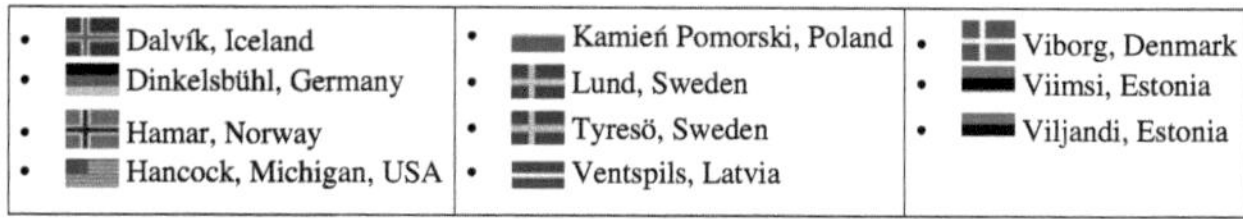

• Dalvík, Iceland	• Kamień Pomorski, Poland	• Viborg, Denmark
• Dinkelsbühl, Germany	• Lund, Sweden	• Viimsi, Estonia
• Hamar, Norway	• Tyresö, Sweden	• Viljandi, Estonia
• Hancock, Michigan, USA	• Ventspils, Latvia	

References

[1] "Area by municipality as of 1 January 2011" (http://www.maanmittauslaitos.fi/sites/default/files/pinta-alat_2011_kunnannimenmukaan. xls) (in Finnish and Swedish) (PDF). Land Survey of Finland. . Retrieved 9 March 2011.

[2] "Population by municipality as of 31 January 2012" (http://vrk.fi/default.aspx?docid=5919&site=3&id=0) (in Finnish and Swedish). *Population Information System*. Population Register Center of Finland. . Retrieved 16 February 2012.

[3] "Population according to language and the number of foreigners and land area km2 by area as of 31 December 2008" (http://pxweb2.stat.fi/ Dialog/varval.asp?ma=060_vaerak_tau_107_fi&ti=Väestö+kielen+mukaan+sekä+ulkomaan+kansalaisten+määrä+ja+ maa-pinta-ala+alueittain++1980+-+2008&path=../Database/StatFin/vrm/vaerak/&lang=3&multilang=fi). *Statistics Finland's PX-Web databases*. Statistics Finland. . Retrieved 29 March 2009.

[4] "Population according to age and gender by area as of 31 December 2008" (http://pxweb2.stat.fi/Dialog/varval. asp?ma=050_vaerak_tau_104_fi&ti=Väestö+iän+(1-v.)+ja+sukupuolen+mukaan+alueittain+1980+-+2008&path=../Database/ StatFin/vrm/vaerak/&lang=3&multilang=fi). *Statistics Finland's PX-Web databases*. Statistics Finland. . Retrieved 28 April 2009.

[5] "List of municipal and parish tax rates in 2011" (http://www.vero.fi/nc/doc/download.asp?id=7996;193801). Tax Administration of Finland. 29 November 2010. . Retrieved 13 March 2011.

[6] http://www.porvoo.fi/

[7] Jaakkola, Marianne (2007-11-19). "Yleistä Porvoosta" (http://matkailu.porvoo.fi/fi/yleista_porvoosta) (in Finnish). Porvoo: City of Porvoo. . Retrieved 7 January 2009.

[8] http://www.katajala.net/keskiaika/suomi/kaupungit.html

[9] http://www.kirkkojakaupunki.fi/uutiset/rikos-ja-rangaistus

[10] "Kerava-Porvoo Museum Train Timetable Summer 2009" (http://www.helsinkiww.net/pmr/eng/time2009.html). Porvoo Museum Railway Society. . Retrieved 14 October 2009.

[11] "The sister cities of Porvoo" (http://www.porvoo.fi/fi/yleistietoa/ystavyyskuntatoiminta) (in Finnish). Porvoo City. . Retrieved 5 March 2011.

External links

- City Porvoo (http://www.porvoo.fi/) – Official website
- Porvoo travel guide from Wikitravel
- King's Gate Business Park (http://www.srv.fi/reference_property_dev?id=1751433)
- Porvoo Museum Railway (http://helsinkiww.net/pmr/index.html) – train service to/from Helsinki on summer Saturdays

Dalvík

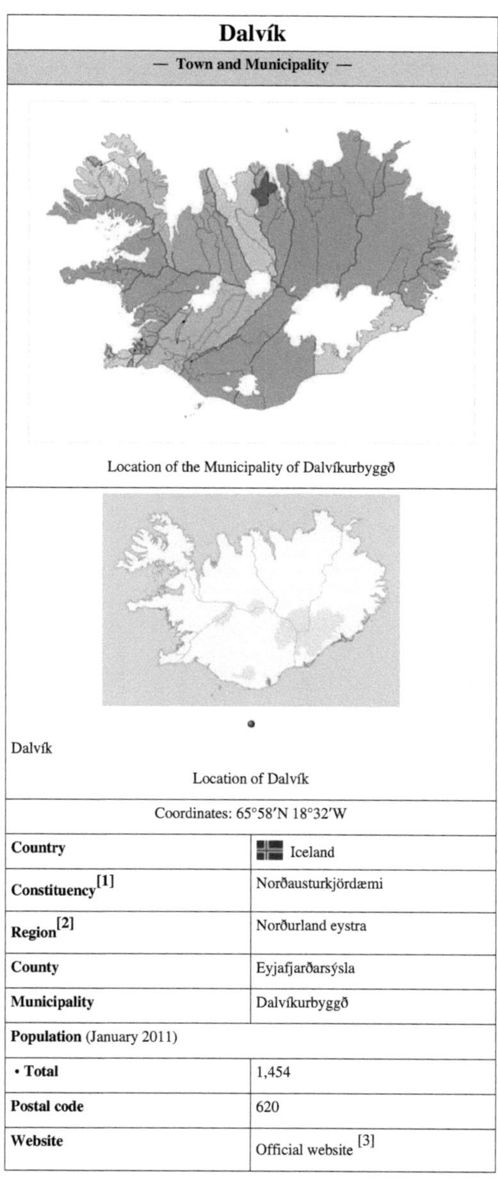

<table>
<tr><td colspan="2" align="center">Dalvík</td></tr>
<tr><td colspan="2" align="center">— Town and Municipality —</td></tr>
<tr><td colspan="2" align="center">Location of the Municipality of Dalvíkurbyggð</td></tr>
<tr><td colspan="2" align="center">Dalvík
Location of Dalvík</td></tr>
<tr><td colspan="2" align="center">Coordinates: 65°58′N 18°32′W</td></tr>
<tr><td>Country</td><td>Iceland</td></tr>
<tr><td>Constituency[1]</td><td>Norðausturkjördæmi</td></tr>
<tr><td>Region[2]</td><td>Norðurland eystra</td></tr>
<tr><td>County</td><td>Eyjafjarðarsýsla</td></tr>
<tr><td>Municipality</td><td>Dalvíkurbyggð</td></tr>
<tr><td colspan="2">Population (January 2011)</td></tr>
<tr><td>• Total</td><td>1,454</td></tr>
<tr><td>Postal code</td><td>620</td></tr>
<tr><td>Website</td><td>Official website [3]</td></tr>
</table>

Dalvík the main village of the Icelandic municipality Dalvíkurbyggð in Iceland.

The population of the village Dalvík is approximately 1,400.[4]

The town's name *Dal-vík* means "dale-bay."

Geography

Dalvík is east of Eyjafjörður in the valley Svarfaðardalur.

Transportation

Dalvík harbor is a regional commercial port for import and fishing. The ferry Sæfari, which sails from Dalvík, serves the island of Grímsey, Iceland's northernmost community, which lies on the Arctic Circle.

Culture

The annual Fiskidagurinn mikli is held the Saturday after the first Monday of August, attended by up to 30,000 people who enjoy a free fish buffet sponsored by the local fishing industry.[5]

Sports

In sports, Dalvík is probably most known for alpine skiing with Böggvisstaðafjall ski area one of the best known ski areas in Iceland. The town has produced a series of skiers that have represented Iceland in the Olympics, World Cups, World Championships, European Cups as well as other international and national competitions. Amongst these have been Daníel Hilmarsson, Sveinn Brynjólfsson and currently Björgvin Björgvinsson.

Football teams from the village have had their ups and downs but have managed to produce some nationally known players but the most recognized one is without a doubt the QPR forward Heiðar Helguson.

Hamar golf club is a thriving club that has a 9 hole golf course a short drive outside Dalvík.

Economy

The local economy is based upon fisheries and fish processing. Dalvík is also a tourist destination for boat trips in whale watching and heli skiing.

References

[1] Political division
[2] Mainly statistical division
[3] http://http://www.dalvik.is/
[4] Hagstofa Íslands (http://www.hagstofa.is/?PageID=624&src=/temp/Dialog/varval.asp?ma=MAN02001&ti=Mannfjöldi+eftir+ sveitarfélagi,+kyni+og+aldri+1.+janúar+1998-2008+++&path=../Database/mannfjoldi/sveitarfelog/&lang=3&units=Fjöldi), *Statistics Iceland* Website
[5] The Great Fish Day (http://www.icelandreview.com/icelandreview/features/multimedia/?ew_news_onlyarea=& ew_news_onlyposition=13&cat_id=29473&ew_13_a_id=287231), *Iceland Review article.*

External links

* Dalvíkurbyggð Municipality (http://www.dalvik.is)
* Fiskidagur - official site (http://fiskidagur.muna.is/)

León,_Nicaragua

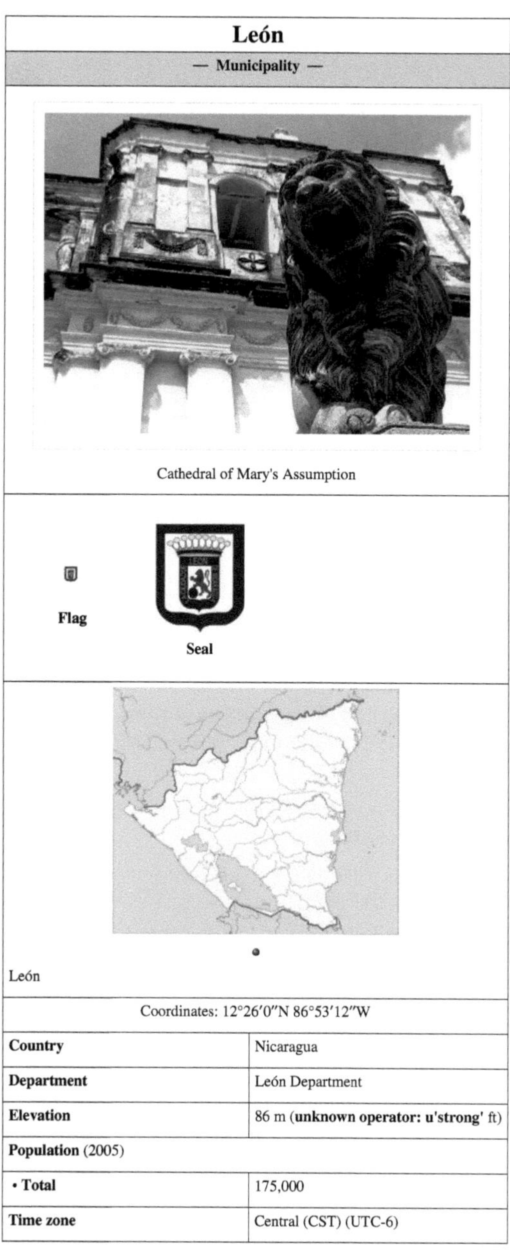

León	
— Municipality —	

Cathedral of Mary's Assumption

Flag

Seal

León

Coordinates: 12°26′0″N 86°53′12″W

Country	Nicaragua
Department	León Department
Elevation	86 m (**unknown operator: u'strong'** ft)
Population (2005)	
• Total	175,000
Time zone	Central (CST) (UTC-6)

León is a department (state) in northwestern Nicaragua (5,138 km^2). It is also the second largest city in Nicaragua, after Managua. It was founded by the Spaniards as **León Santiago de los Caballeros** and rivals Granada, Nicaragua, in the number of historic Spanish colonial homes and churches. As of 2005, the city had an estimated population of about 175,000 people which increases sharply during university season with many students coming from other Nicaraguan provinces. It is the capital and municipality of the León department.

León is located along the Río Chiquito *(Chiquito River)*, some 90 km northwest of Managua, and some 18 km east of the Pacific Ocean coast. Although less populous than Managua, León has long been the intellectual center of the nation, with its university founded in 1813. León is also an important industrial, agricultural (sugar cane, cattle, peanut, plantain, sorghum) and commercial center for Nicaragua.

History

The first city named León in Nicaragua was established in 1524 by Francisco Hernández de Córdoba about 20 miles east of the present site. The city was abandoned in 1610, for unknown reasons. The principal cause is commonly given as a necessary abandonment after an eruption of the Momotombo volcano, located only a couple miles away, which left extensive damage in the form of flooding from Lake Managua. However, the speed of the construction of the new León suggest that the old city was in great part dismantled, moved, and rebuilt, and therefore must have happened before the destruction of the site by the volcano. Other possible reasons for the move include the need for fresh agricultural land, the need for higher concentrations of natives to use as a labour force, and perhaps also fear of Momotombo erupting - although unrecorded, it could have been releasing gas, ash, or other volcanic material for some time before the eventual eruption. The inhabitants decided to move to its current location next to the Indigenous town of *Subtiava*. The ruins of the abandoned city are known as "León Viejo" and were excavated in 1960. In the year 2000, León Viejo was declared a UNESCO World Heritage Site.

León has fine examples of Spanish Colonial architecture, including the grand Cathedral of the Assumption, built from 1706 to 1740, with two towers added in 1746 and 1779. In the year 2011, the Cathedral of the Assumption was declared a UNESCO World Heritage Site.

When Nicaragua withdrew from the United Provinces of Central America in 1839, León became the capital of the new nation of Nicaragua. For some years the capital shifted back and forth between León and Granada, Nicaragua, with Liberal regimes preferring León and Conservative ones Granada, until as a compromise Managua was agreed upon to be the permanent capital in 1858.

In 1950 the city of León had a population of 31,000 people. Nicaraguan President Anastasio Somoza García was shot and mortally wounded in the city on September 21, 1956.

The building of El museo de tradiciones y leyendas was once the infamous XXI jail before the 1979 revolution. There are also several political murals around the city.

Heritage

The heritage of León is rich. Both monuments and natural places in response to the monuments include the following: Cathedral Basilica of the Assumption, León, 2007.

- **Cathedral Basilica of the Assumption of León**, typical colonial baroque building was built between 1747 and 1814. Because strength of its walls has endured earthquakes, volcanic eruptions of the volcano Cerro Negro and wars. In 1824, were placed several cannons on the roof during the siege of the city by conservative forces; and, in the uprising of June and July 1979 against the dictator Anastasio Somoza Debayle, the guerrillas of the Sandinista National Liberation Front also used for war purposes.

    This cathedral is one of the largest in Central America. It was the first episcopal seat of Nicaragua, since 1531, making it one of the oldest dioceses in the Americas. It is the tomb of the poet Rubén Darío, at the foot of the statue of St. Paul, leading figure of modernism and considered the *Prince of the Castilian literature*. In its crypts, designed to withstand earthquakes, are buried some illustrious people of the nation as Salomón de la Selva and Alfonso Cortés, the hero Miguel Larreynaga and musician José de la Cruz Mena.

There are a number of tunnels that connect this church with other churches in León. In the early 20th century the first bishop of León and last in Nicaragua, Archbishop Simeón Pereira y Castellón (the same who presided over the funerals of Darío on 13 February 1916) commissioned the Granadan sculptor Jorge Navas Cordonero make the statue of the Virgin Mary above the front of the facade, the Atlanteans that are among the gables and the towers. Navas also sculpted the statues of the Twelve Apostles, along with the columns of the central nave, like the lion of the tomb of the poet, much look like at the Lion of Lucerne, Switzerland, and various decorations inside the church and its Tabernacle Chapel.

- **Church of Subtiava** was considered the main church after the cathedral. Its construction began in 1698, at time of magistrate Diego Rodríguez Menéndez and was completed August 24, 1710. In the war with El Salvador, in 1844, disappeared the dome that crowned the tower so was re-built in the early 20th century again.
- **Church of San Francisco** is part of the convent of San Francisco, one of the oldest in Nicaragua, founded in 1639 by Friar Pedro de Zúñiga. The interior remains two good examples of plateresque altars.
- **Church of La Recolección**, construction began on December 5, 1786, by Bishop Juan Félix de Villegas. made thanks to the gatherer fathers of congregation San Francisco de Nery. It is a good baroque facade (of Mexican Baroque), considered the most important of the city. Its altar, also Baroque, is one of the best altarpieces of the city. Highlights the paintings and engravings with which it adorns.
- **Church La Merced**, in 1762 the Mercedarian fathers built their convent and the church that were demolished. In the 18th century was rise the present Church of La Merced with drawings attributed to Mercedario Friar Pedro de Ávila and conducted by master Pascual Somarriba. Adjacent to the north side is the building del Paraninfo (former mercedario convent), main building of the UNAN-León.
- **Church El Calvario**, interesting town-planning for its spatial distribution, this church dates from the first half of the 18th century but was modified the north tower in the 20th century. It was built by the illustrious Mayorga Family.
- **Ruins of the Church of San Sebastián**, built in the late 17th century as a chapel of the Cathedral and was one of the first religious buildings of the city. Re-built in late 18th century by Colonel Joaquín Arrechavala. It was bombed during the siege of León by airplanes of Nicaraguan Air Force in 1979, due was made of adobe was easily destroyed unlike other churches, which, being built of brick and stone quarry, endured the fighting.
- **Church Nuestra Señora de Guadalupe**, built in the late 19th century under the auspices of father Villamil replacing an hermitage of 18th century, is of simple construction, in keeping with the sobriety of the Franciscans.
- **Church of Zaragoza**, with an atrium and lateral corridor, its construction began in the late 19th century and ended in mid-20th century by Bishop Salmerón, the facade was designed by Dr. Francisco Mateo.
- **Church San Felipe**, a large building that occupies an entire block, was built in 1685 for blacks and mulattos can pray. In 1859 undergoes extensive expansion that gives the present form, the tower was restored in 1983.
- **Church Hermitage of San Pedro**, a small building with typical popular architecture of the 18th century. It was built between 1706 and 1718 by the mayor Bartolomé González Fitoria, replacing the original church of San Pedro that was part of a set of four primitive hermitages in Subtiava.
- **Church San Nicolás de Tolentino del Laborío**, Philip III of Spain ordered its construction in 1618. Of Colonial baroque style, is of lines very light.
- **Chapel de la Asunción**, part of the school of the same name and was built in 1679 by Bishop Andrés de las Navas y Quevedo, used as an episcopal palace. Later it was occupied by the Mothers of the Assumption, and in 1935, were made deep reforms that led to the current appearance.
- **Hotel Esfinge** building designed by Nicaraguan architect José María Ibarra like upscale hotel.

- **Esquivel House**, Built in the 19th century belonged to Father Mariano Dubon 20th century around the house detail and decoration of the structure makes it unique in this city the house now belongs to the **Esquivel Family** and this is categorized and rated as high-class residence.

- **Colegio San Ramón**, of symmetrical facade and Renaissance influences has been raised several times due was destroyed by earthquakes. Has been occupied by the university and has had seminar roles.

- **City Hall**, built in 1935, was damaged in 1979 by Somoza's repression of the National Guard against the rise of the FSLN. The building was designed by architect Marcelo Targa, precursor of neoclassical architecture in Leon, following the artistic movement of the late 20th century. It was built during the administration of Juan Bautista Sacasa. It is a building of important architectural value.

- **León Viejo**, the ruins of the ancient city of León which was buried by eruptions of the volcano Momotombo, most of them in 1610. Was founded in 1524 by Francisco Hernández de Córdoba, it is situated 30 km from the modern city, is a World Heritage Site and in this is buried its founder, in a crypt, beneath his statue along with other characters including Pedrarias Dávila his murderer.

There are more buildings and important places, like the walls of the cemetery of Guadalupe, the Guadalupe Bridge, Sutiava Rural House or the train station.

Sesteo Restaurant near Cathedral and Main Square.

Cathedral Basilica of the Assumption of León.

Church El Calvario.

Spanish colonial home.

Spanish colonial home.

Couple at Poneloya Beach, just outside León.

Poneloya Beach, just outside Leon at sunset.

Church of La Recolección

Geological features around Leon include:

- **Poneloya beach**, a major tourist destination on the Pacific Ocean.

- **San Jacinto Swarms**, geothermal place at the base of the Santa Clara volcano and constitute part of its vents. This phenomenon is known as *Thermal sharpen* to simmer.

- **Momotombo volcano**, with its 1300 m summit, it is the reference to Leonese visual landscape. At its base lies *León Viejo*. The name is a Native American word for *Great Summit Burning*. A geothermal power generation exists onsite.

- **Cerro Negro volcano**, one of the youngest of the earth (1850), has caused trouble for the population of León through its ashfall in recent eruptions.

Notable people

- José de la Cruz Mena, greatest Nicaraguan classical composer (1874–1907).
- Azarías Pallais, one of Nicaragua's greatest poets and literary figures (1884–1954).
- Alfonso Cortés, most renowned Nicaraguan poet after Rubén Darío (1893–1969).
- Salomón de la Selva, poet, writer, diplomatic, translated W Whitman to Spanish (1893–1959).
- Antenor Sandino, greatest metaphysical poet (1899–1969).
- Anastasio Somoza Debayle, Nicaraguan president (1st term: 1967–1972, 2nd term: 1974–1979); overthrown by the Sandinistas.
- Enrique Bermúdez, Nicaraguan Contra leader (1932–1991).

Sister cities

- Hamburg, Germany
- Oxford, United Kingdom
- Gettysburg, Pennsylvania, USA
- New Haven, Connecticut, USA
- Utrecht, Netherlands
- Lund, Sweden
- Alicante, Spain
- Janesville, Wisconsin, USA

See also

- León Viejo (archaeological site)

External links

- Leon: hotels, restaurants, activities, photos [1]

References

[1] http://www.vianica.com/visit/leon

Article Sources and Contributors

Urban_areas_in_Sweden *Source*: http://en.wikipedia.org/w/index.php?title=Urban_areas_in_Sweden *Contributors*: BD2412, Bbx, CalJW, Danog-76, David ekstrand, Dewritech, Erik031, Everyking, Family Olofsson, Fred J, Gaius Cornelius, Gerrit, Hardouin, Havreflan, Jensapag, Johan Magnus, Joriki, Karmosin, Koyos, LA2, Materialscientist, Muniswede, NULL, Once in a Blue Moon, Pais, Paulleake, Pia L, RedWolf, Rjwilmsi, Ruhrjung, Sam Hocevar, Skizzik, Skrofler, TOCS2011, Thehelpfulone, Thomas Blomberg, Tkynerd, Tomas e, TopoCode, Torsten22, მაჭავარიანი, 26 anonymous edits

Lund_Municipality *Source*: http://en.wikipedia.org/w/index.php?title=Lund_Municipality *Contributors*: .:Ajvol:., BjörnBergman, BrianHansen, Bvalltu, Css, DerBorg, Family Olofsson, Finlay McWalter, Fred J, Fredrik, Gauss, Gunnar Larsson, Hankwang, Hede2000, Jensapag, Jeronimo, Johan Magnus, Jonkerz, LA2, Lavallen, Liftarn, Mahanga, Mic, Muniswede, Naddy, Paul Richter, Per Hedetun, Plastikspork, Skizzik, Stern, Sverdrup, Swirl en, Tarquin, Template namespace initialisation script, TerriersFan, The monkeyhate, Tomas e, Tuomas, Uppland, 26 anonymous edits

Municipalities_of_Sweden *Source*: http://en.wikipedia.org/w/index.php?title=Municipalities_of_Sweden *Contributors*: Acntx, Alarm, Andhanq, Andycjp, Barticus88, Beland, Bjankuloski06en, Can't sleep, clown will eat me, Cassowary, Chanheigeorge, Charlierules1, Cmdrjameson, CodeGeneratR, Crusoe8181, D6, Dario Zornija, David ekstrand, Docu, Earl Andrew, Ecspalt, Enlightener, Essin, Euchiasmus, Family Olofsson, Fred J, Funandtrvl, Gaius Cornelius, Gamle Bailey, Hmains, John Vandenberg, Lightmouse, Lokal Profil, M.M.S., MBisanz, Marek69, Mic, Mrom, Muniswede, Mywood, Roke, Røed, Saihtam, Sardanaphalus, Slarre, SwahilidR, Template namespace initialisation script, TimSE, Tobias Conradi, Tolanor, Vikingviolinist, Wik, Wikid77, Wilhelm meis, 24 anonymous edits

Skåne_County *Source*: http://en.wikipedia.org/w/index.php?title=Sk%C3%A5ne_County *Contributors*: *Ulla*, Ahoerstemeier, Altzinn, Aron Boström, Berkay0652, Bilar-, Bjankuloski06en, BjörnBergman, Borgx, Caponer, Cattus, Commander Keane, D6, Delirium, Derek Ross, Egil, Fred J, Gaius Cornelius, Hayden120, Heymid, Iwillremembermypassthistime, JT Swe, Jao, Jauhienij, Jeronimo, Johan Elisson, John Anderson, Jonkerz, Kaiser Torikka, Liftarn, MapsMan, Marsaskala, Mic, Mintleaf, Mr. Atom Scania, Muniswede, Nolose, Numbo3, Orioane, Oxxo, Pavao Zornija, Pegship, Ph.eyes, Pia L, Reckless182, Rich Farmbrough, Robertgreer, Ruhrjung, Sardanaphalus, Skizzik, Slovolyub, Svennis, Sverdrup, TERdON, Template namespace initialisation script, Tfine80, The Great Cucumber, Tomas e, Vedum, Wik, Wilhelm meis, Wisg, Yzmo, მაჭავარიანი, 31 anonymous edits

Dalby_Söderskog_National_Park *Source*: http://en.wikipedia.org/w/index.php?title=Dalby_S%C3%B6derskog_National_Park *Contributors*: Aranel, Beckmans, Bvalltu, Clarkbhm, Darwinek, Droll, Family Olofsson, Fred J, Jaraalbe, Johan Elisson, Jonkerz, Kdhenrik, MIKHEIL, Mic, Pegship, Ph.eyes, Scaune, Skizzik, Sverdrup, Theleftorium, Tomas e, Väsk, Williamborg, 6 anonymous edits

List_of_national_parks_of_Sweden *Source*: http://en.wikipedia.org/w/index.php?title=List_of_national_parks_of_Sweden *Contributors*: Angela, Bjankuloski06en, Bobblewik, Bryan Derksen, Bvalltu, Chrishmt0423, ClaudeMuncey, CommonsDelinker, Crusoe8181, Dabomb87, David ekstrand, Eubulides, Fred J, Grapetonix, Hu12, Inwind, J.delanoy, Kaare, Kumioko (renamed), Mic, Neelix, Pizza Puzzle, Ppntori, RexxS, Reywas92, Rich Farmbrough, Ruslik0, Skizzik, The Rambling Man, Theleftorium, Tournesol, Unyoyega, Vanjagenije, WaitingForConnection, Williamborg, 12 anonymous edits

Dalby,_Lund *Source*: http://en.wikipedia.org/w/index.php?title=Dalby%2C_Lund *Contributors*: Amorymeltzer, D6, Fred J, Grillo, Grutness, Henrygb, His047, Jensapag, Johan Magnus, Karmosin, M3926-990031, Muniswede, Olaus, Pegship, Petri Krohn, Pia L, Rogerfletcher, Seb8808, Skizzik, Srnec, Stemonitis, Sverdrup, Väsk, Warfvinge, 7 anonymous edits

Viborg_Municipality *Source*: http://en.wikipedia.org/w/index.php?title=Viborg_Municipality *Contributors*: Agathoclea, Angelbo, Borgx, Bornsommer, DS1953, DepartedUser4, DerBorg, Dycedarg, Lilac Soul, Mimihitam, Muniswede, Numbo3, Pixi Uno, Sfdan, Tim!, TopoCode, Twthmoses, Valentinian, Wavelength, 4 anonymous edits

Hamar *Source*: http://en.wikipedia.org/w/index.php?title=Hamar *Contributors*: -xfi-, Achangeisasgoodasa, Ahoerstemeier, Andres, AnnaFrance, Aqwis, Arsenikk, Arthena, BigNate37, Bonadea, Charles Matthews, Circeus, Clq, Conorbrady.ie, Conversion script, Cxw, Cyppen, DBrane, David Schaich, Drbreznjev, Dwo, Egil, Ehe93, Eric82oslo, Fastifex, Fgfgter, Fredrik, GalFisk, Geschichte, Gilliam, Griffindd, Hmains, Houshuang, Ingeha, Ingvarson39, Iridescent, J04n, Jackie, Jauhienij, Jay1279, Jeff no, Jeronimo, Jevansen, Jhendin, John Carter, Jon Harald Søby, Jorunn, Jyusin, Korovioff, Kummi, Leifern, Magioladitis, Marek69, Middayexpress, Naitsirk, Nissen3214, Nlu, Numbo3, Oeuftete88, Olasveengen, Optigan, PBS-AWB, Petri Krohn, Reedy, Rettetast, Rich Farmbrough, Rjwilmsi, Rookkey, Rwk1, Sabergum, Sagaciousuk, Sam Vimes, Samogitia, Samuelsen, Silvonen, Sjakkalle, Skysmith, Spelemann, Tabletop, Tassedethe, Template namespace initialisation script, Torstein, Uppland, Wackywace, Warfvinge, WhisperToMe, Williamborg, Žiga, 77 anonymous edits

Porvoo *Source*: http://en.wikipedia.org/w/index.php?title=Porvoo *Contributors*: AB, Aaker, Ahoerstemeier, Andre Engels, Anthony aragorn, Apalsola, Archtransit, Benediktee, Caid Raspa, Coemgenus, Colonies Chris, Darklilac, Dekimasu, Den fjättrade ankan, Easyas12c, Enchanter, Esprit15d, Fannymalin, Flying Saucer, Gaius Cornelius, Gnomeselby, Grafikm fr, Howcheng, Hughcharlesparker, Jaakko Sivonen, Janke, JdeJ, Jhendin, Joao Xavier, Jonik, Joseph Solis in Australia, Jpatokal, Jyril, Koavf, Kotasik, Kwamikagami, Leopea, Lugnuts, Mahmudss, Mangostar, Marek69, Matthewross, Mic, Monegasque, Morwen, Multixfer, Nina Smith, Nn123645, Ollisiren, Otkdkr, Oy Maatsulu, Pairadox, Pallojalka, Pavel Vozenilek, Pearle, Piechjo, Ptk, Reimgild, Remi1992, Rich Farmbrough, Roton6, RuED, Ruhrjung, Russophile2, Samulili, Silvonen, Simonkoldyk, Stadscykel, Stemonitis, SteveDay, Suedois, Surfo, TTKK, Template namespace initialisation script, Thierry Caro, Thrór, Tonym88, Ultrix, Wasted Time R, Wesley, XxTimberlakexx, Zakuragi, 100 anonymous edits

Dalvík *Source*: http://en.wikipedia.org/w/index.php?title=Dalv%C3%ADk *Contributors*: Debivort, Gruesome Gary, John, Jón, 3 anonymous edits

León,_Nicaragua *Source*: http://en.wikipedia.org/w/index.php?title=Le%C3%B3n%2C_Nicaragua *Contributors*: Alansohn, Amikake3, Amire80, Angela, Angelo De La Paz, Attilios, Auslli, Bletch, Boomtown Rat, Chris Roy, Colonies Chris, Darwinek, DiRkdARyL, Dp462090, Dr. Blofeld, Eafb, Edivorce, EncMstr, Everyking, FromManagua, Gene Nygaard, Greenshed, Hacostaj, Hajor, Hanek45, Herman Downs, Highground79, Infrogmation, Jalo, Joao Xavier, L1A1 FAL, LaNicoya, Leontour2011, Lotje, Man Usk, Marek69, MarritzN, Martinete, Mikkelanjelo, MinnesotanConfederacy, Mysekurity, Nappyrootslistener, Nicapa, PMDrive1061, Pb30, Pgk, Qrfqr, Quarty, Qxz, Rastrojo, Raymond, Ronflacayo, Ryos, Sam Hocevar, Sardanaphalus, SchuminWeb, Serenityweb, Serenityweb1, StAnselm, StefanB sv, Stepheng3, Tibetan Prayer, Unyoyega, Venerock, Vrysxy, Whagers, Wik, Wjddbsals, X201, ZS, °, 72 anonymous edits

Image Sources, Licenses and Contributors

Printed by Books on Demand GmbH, Norderstedt / Germany